DIY Electronics Repair at Home

Chapters

g) **Manufacturing and Industry**: Electronics are integral to modern manufacturing processes. Automation, robotics, and sensors improve productivity, quality control, and safety in factories.

h) **Environmental Monitoring**: Electronics play a crucial role in environmental monitoring and conservation efforts. They enable data collection, analysis, and real-time reporting, helping address climate change, pollution, and natural resource management.

i) **Space Exploration**: Electronics are vital in space exploration. They power spacecraft, collect data from distant planets, and enable communication with astronauts on space missions.

j) **National Security**: Electronics support national security through advanced surveillance, encryption, and defense technologies. They play a critical role in safeguarding nations from external threats.

k) **Research and Development**: Electronics accelerate scientific research, allowing scientists to conduct experiments, model complex systems, and analyze data more efficiently.

l) **Agriculture**: Electronics support precision agriculture by enabling remote monitoring of crops, soil conditions, and livestock, leading to improved yield and resource utilization.

m) **Smart Cities**: Electronics are integral to the development of

Chapter 1: Introduction to Home Electronics Repair

1. Highlight the Relevance of Electronics

Electronics play a pivotal role in various aspects of our modern world, offering significant relevance in many areas of our lives. Here are some key points highlighting the relevance of electronics:

a) **Communication**: Electronics enable instant global communication through devices like smartphones, computers, and the internet. They connect people smart cities. They support systems for traffic management, waste disposal, energy conservation, and public safety.

n) **Banking and Finance**: Electronics enable secure and efficient financial transactions, online banking, and electronic payment systems, improving convenience for consumers and businesses.

o) **Education**: Electronics enhance the educational experience through e-learning platforms, digital textbooks, and interactive classroom technology, making education more accessible and engaging.

2. Benefits and Challenges of Home Electronics Repair

- Benefits:

i) **Cost Savings**: One of the primary benefits of home electronics repair is cost savings. Repairing your electronics and appliances can be more affordable than replacing them with new ones. It can extend the life of your devices, saving you money in the long run.

ii) **Convenience**: You have control over the repair process when doing it at home. You can work on the repair at your own pace and schedule, eliminating the need to transport the device to a repair shop or wait for a technician's visit.

iii) **Skill Development**: Home electronics repair provides an opportunity to learn new skills and gain hands-on experience in electronics troubleshooting and repair. It can be a rewarding and educational hobby or even a potential career path.

iv) **Environmental Impact**: Repairing electronics and appliances reduces electronic waste. This is environmentally responsible and contributes to sustainability efforts by decreasing the disposal of e-waste in landfills.

v) **Self-Sufficiency**: The ability to repair your own electronics and appliances promotes self-sufficiency and empowers you to take control of your possessions.

You're less reliant on external repair services.

- Challenges:

i) **Safety Risks**: Electronics repair can be dangerous if not done properly. Some devices, especially those with high voltage components, can pose electrical and fire hazards. Safety precautions are crucial.

ii) **Warranty Issues**: Attempting repairs on devices that are under warranty may void the warranty. It's essential to consider this before deciding to repair the device at home.

iii) **Complexity**: Some electronics are highly complex, with intricate components and proprietary designs. Diagnosing and repairing these devices can be challenging, and specialized tools may be necessary.

iv) **Lack of Resources**: Repairing modern electronics often requires specialized tools, service manuals, and replacement parts. Access to these resources can be limited for certain devices.

v) **Time-Consuming**: Repairing electronics and appliances can be time-consuming, especially if you're not experienced. Some repairs may require a significant time investment.

vi) **Frustration**: Home repair attempts can lead to frustration if you encounter unexpected difficulties or if the repair is unsuccessful. In some cases, it may be more practical to seek professional help.

vii) **Risk of Further Damage**: Inexperienced repair attempts can sometimes worsen the problem or cause additional damage to the device, leading to higher repair costs or rendering the device irreparable.

3. Safety Precautions and Considerations

When undertaking home electronics repair, it's essential to prioritize safety, both for yourself and the proper functioning of the device. Here are some key safety precautions and considerations to keep in mind:

a) **Disconnect the Power**: Always unplug the device or, if applicable, turn off the power source before beginning any repair. This is the first and most crucial safety step.

b) **Wear Safety Gear**: Depending on the repair and the specific device, it may be wise to wear safety gear such as safety glasses, anti-static wristbands (if working on sensitive electronics), or heat-resistant gloves.

c) **Work in a Clean Environment**: A clean and organized workspace helps prevent accidents and misplaced tools. Avoid clutter, which can lead to misplaced components or tools.

d) **Use the Right Tools**: Ensure you have the appropriate tools for the job. Using the wrong tools can lead to damage, injury, or inaccuracies in the repair. Some tools may be specific to certain devices.

e) **Educate Yourself**: Understand the device and its components before starting any repair. Study service manuals, repair guides, and online resources related to the device.

f) **Avoid Water and Liquids**: Electronics and water do not mix. Keep liquids away from your workspace and ensure your hands are dry.

g) **Anti-Static Precautions**: When working on sensitive electronic components, ground yourself to prevent static electricity discharge, which can damage components. Use anti-static mats and wristbands.

h) **Work on a Non-Conductive Surface**: Place the device on a non-conductive surface, such as an anti-static mat or a wooden table, to prevent electrical shorts.

i) **Check for Capacitor Discharge**: Some devices store electrical energy in capacitors even when unplugged. Discharge capacitors safely using a discharge tool or resistor if necessary.

j) **Proper Ventilation**: When soldering or working with adhesives, ensure you have adequate ventilation to prevent inhaling harmful fumes.

k) **Label and Organize**: Label cables, connectors, and screws to keep track of their positions during disassembly. This will make reassembly easier.

l) **Be Cautious with High Voltage Devices**: Devices with high-voltage components (e.g., CRT monitors, some power supplies) can be extremely dangerous. Avoid contact with high-voltage components and discharge capacitors.

m) **Don't Force It**: If a component or part does not come apart easily, don't force it. Forcing components can lead to breakage or damage.

n) **Keep Children and Pets Away**: Ensure that children and pets are not in the workspace to prevent accidents or injuries.

o) **Know When to Stop**: If you encounter a problem or issue beyond your expertise or if you become frustrated, it's often better to seek professional assistance rather than risk further damage.

p) **First Aid Kit**: Have a basic first-aid kit on hand in case of minor injuries.

q) **Fire Safety**: Be prepared for a fire emergency. Have a fire extinguisher nearby, and know how to use it.

r) **Proper Disposal**: Dispose of damaged or unusable components and hazardous materials, like old batteries, in accordance with local regulations.

Chapter 2: Essential Tools and Equipment

1. Fundamental Tools

Performing home electronics repair effectively and safely requires a set of fundamental tools and equipment. Here is a list of essential tools for home electronics repair, along with brief descriptions of their uses:

a) **Screwdrivers**: A variety of screwdrivers, both flat-head and Phillips, in different sizes are essential for opening and securing components and cases.

b) **Needle-Nose Pliers**: These allow you to grip and manipulate small components, bend wires, and reach tight spaces.

c) **Wire Cutters and Strippers**: Used to cut and strip wire insulation for soldering and reconnection.

d) **Soldering Iron and Solder**: A soldering iron is crucial for soldering and desoldering electronic components, wires, and connections.

e) **Desoldering Pump or Solder Wick**: These tools are used to remove solder when replacing components or making repairs.

f) **Anti-Static Wristband and Mat**: These tools help prevent static electricity discharge, which can damage sensitive electronic components.

g) **Multimeter**: A digital multimeter measures voltage, current, and resistance. It's invaluable for diagnosing electrical issues and checking component functionality.

h) **Tweezers**: Precision tweezers are helpful for handling small components, aligning tiny parts, and placing them accurately.

i) **IC Extraction Tool**: Used to safely remove integrated circuits (ICs) from their sockets without damaging the pins.

j) **Precision Screwdriver Set**: For working on small, delicate components like laptops, cameras, and smartphones.

k) **Magnifying Glass or Loupe**: These aids in examining small, intricate components and solder joints.

l) **Brushes and Compressed Air**: Used to clean dust, debris, and dirt from electronic components and boards.

m) **Heat Gun**: Useful for reflowing solder or removing adhesives during certain repairs.

n) **Spudgers and Pry Tools**: These non-conductive tools help open cases, separate components, and remove adhesives without causing damage.

o) **Safety Gear**: Safety glasses, heat-resistant gloves, and an anti-static smock or mat provide personal protection during repairs.

p) **Small Parts Containers**: To organize and store screws, connectors, and other small components when disassembling devices.

q) **Adhesive and Tape**: Used to reattach components and secure cables or wires.

r) **Anti-Static Bags**: Useful for storing sensitive components or electronics to protect them from static electricity.

s) **First-Aid Kit**: A basic first-aid kit should be on hand in case of minor injuries.

t) **Storage and Organization**: Toolboxes or organizers help keep your tools neat, accessible, and in good condition.

2. Tool Selection and Maintenance

- Tool Selection:

a) **Quality Tools**: Invest in high-quality tools from reputable manufacturers. Quality tools are more durable and provide better results. They are less likely to damage components or strip screws.

b) **Tool Kit**: Consider purchasing a comprehensive tool kit designed for electronics repair. These kits often include a variety of essential tools in one package.

c) **Specialized Tools**: Depending on the types of repairs you plan to undertake, you may need specialized tools like a hot air rework station for SMD components or a logic analyzer for digital circuit testing.

d) **Ergonomics**: Choose tools with comfortable handles and ergonomic designs. You may be using them for extended periods, and comfort can reduce fatigue and improve precision.

e) **Anti-Static Tools**: For working with sensitive electronics, invest in anti-static versions of tools, including wristbands, mats, and brushes.

• Tool Maintenance:

a) **Clean Tools**: After each use, clean your tools with a soft, dry cloth or a brush to remove debris, dust, or residue. For stubborn residues, use isopropyl alcohol or a specialized electronics cleaning solution.

b) **Keep Tools Organized**: Store your tools in an organized manner to prevent damage and make them easy to find when needed. Toolboxes, organizers, or pegboards can help with this.

c) **Regular Calibration**: For tools like multimeters or oscilloscopes, follow manufacturer guidelines for calibration and regular accuracy checks.

d) **Replace Worn-Out Parts**: If you notice wear and tear, replace parts like worn screwdriver tips or soldering iron tips. Using worn tools can lead to poor results and even safety hazards.

e) **Lubrication**: For tools like pliers or wire cutters, apply a small amount of lubricant to moving parts to prevent rust and ensure smooth operation.

f) **Soldering Iron Maintenance**: For soldering irons, regularly clean the tip with a sponge or brass wool to remove oxidation. Apply a thin layer of solder to the tip before turning it off to prevent oxidation during storage.

g) **Inspect Electrical Cords**: Inspect power cords and cables for damage or fraying. Replace them if you notice any issues to avoid electrical hazards.

h) **Replace Consumables**: Consumables like solder, solder wick, and desoldering pumps have a limited lifespan. Replace them as needed to maintain effective soldering and desoldering.

i) **Safety Checks**: Check your safety gear, such as safety glasses and gloves, for any damage or signs of wear. Replace them if necessary.

j) **Proper Storage**: Store your tools in a clean, dry, and climate-controlled environment to prevent rust and corrosion.

k) **Scheduled Maintenance**: Consider a scheduled maintenance routine where you inspect, clean, and maintain your tools at regular intervals.

3. Safety Equipment and Protective Gear

Safety equipment and protective gear are essential for ensuring your well-being during home electronics repair. Here's a list of items you should consider having on hand:

a) **Safety Glasses**: Protect your eyes from debris, solder splatters, and small component parts that can fly during repair work.

b) **Heat-Resistant Gloves**: These are useful when working with soldering irons, heat guns, or other tools that generate heat.

c) **Anti-Static Wrist Strap**: Prevent electrostatic discharge (ESD) by grounding yourself when working with sensitive electronics components, such as integrated circuits and computer motherboards.

d) **Anti-Static Mat**: Lay down an anti-static mat on your workspace to provide a controlled surface that dissipates static electricity and prevents ESD.

e) Fume Extractor or Ventilation: When soldering or using adhesives, a fume extractor or good ventilation is necessary to remove potentially harmful fumes and airborne particulates.

f) Respirator or Dust Mask: For tasks involving significant dust or fumes, a respirator or dust mask can protect your respiratory system.

g) First-Aid Kit: Have a basic first-aid kit on hand to address minor injuries like cuts, burns, or scrapes.

h) Fire Extinguisher: Keep a fire extinguisher nearby for immediate response to small fires that may occur during soldering or electrical work.

i) ESD-Safe Clothing: In addition to an anti-static wrist strap and mat, wearing ESD-safe clothing can help minimize electrostatic discharge.

j) Chemical-Resistant Gloves: For handling certain chemicals, adhesives, or cleaning solutions, chemical-resistant gloves provide protection against skin contact.

k) Ear Protection: When working with loud tools or equipment, like some power tools or heat guns, use ear protection to prevent hearing damage.

l) **Steel-Toed Shoes**: In environments where heavy equipment or tools might be moved, steel-toed shoes provide foot protection.

m) **Non-Slip Footwear**: Ensure your footwear provides good traction to prevent slips, especially when dealing with spilled liquids or working on uneven surfaces.

n) **Face Shield or Goggles**: For tasks involving high-speed rotary tools, grinding, or other activities where eye protection is critical.

o) **Apron or Smock**: Protect your clothing from solder splatters, adhesive drips, or other repair-related spills.

p) Voltage Detector: This device can help you identify live electrical circuits and prevent accidental shocks.

q) Lockout/Tagout Equipment: When working on electrical devices, lockout/tagout equipment is crucial to safely isolate power sources and prevent unexpected re-energizing.

r) Safety Poster or Reference: Display a safety poster or reference guide that provides information on best practices and emergency procedures in your workspace.

4. Organizational Tips

Efficient organization is key to successful home electronics repair. Proper organization can help you save time, reduce frustration, and prevent accidental damage. Here are some organizational tips for home electronics repair:

a) Workspace Setup:

➤ Designate a dedicated workspace for electronics repair. A clean, well-lit, and organized area is ideal.

➤ Use an anti-static mat on your work surface to protect sensitive components from electrostatic discharge.

➤ Ensure easy access to power outlets for soldering irons and other tools.

b) Tool Organization:

➢ Keep your tools organized and within reach. Use toolboxes, pegboards, or wall-mounted tool organizers.
➢ Label or mark the spots for each tool to make it easier to find and return them to their proper places.
➢ Store small components, such as screws, in labeled containers or small bins.

c) Safety Equipment:

➢ Have safety gear, such as safety glasses, anti-static wristbands, and heat-resistant gloves, readily available in your workspace.

d) Parts and Components:

➤ Use small containers or organizers with compartments to keep electronic components like resistors, capacitors, and ICs organized.
➤ Label each container with the component's name, value, and part number for easy identification.
➤ Keep an inventory of common components to reduce the need for last-minute trips to the store.

e) Workspace Cleanliness:

➤ Maintain a clean and clutter-free workspace to prevent accidental damage or loss of small parts.

➢ Use brushes and compressed air to remove dust and debris from circuit boards and components.

f) Documentation:

➢ Refer to service manuals, schematics, and repair guides as needed. Have these documents readily accessible on a nearby shelf or in digital form on your computer.

g) Cable Management:

➢ Use cable ties or Velcro straps to manage cables and prevent them from tangling or getting in your way.

h) Testing and Diagnostic Equipment:

➤ Keep diagnostic tools, such as a multimeter, oscilloscope, or logic probe, well-organized and easily accessible.

i) Disposable Materials:

➤ Have a designated area for disposable materials like used soldering iron tips, soldering wick, and cleaning materials.

j) Personal Workspace Rules:

➤ Establish rules for your workspace to ensure that everyone in your household understands the importance of organization and safety.

k) Safety and Emergency Procedures:

➤ Display a safety poster or reference guide that includes emergency procedures and contact information in case of accidents.

l) Storage:

➤ Use sealable plastic bags or anti-static bags to store partially disassembled devices or components to prevent loss or contamination.

m) Organization System:

➢ Develop a labeling system for cables, connectors, and screws to ease reassembly.
➢ Keep an organized digital record of your past repairs, including photos, schematics, and notes, to aid in future repairs.

n) Waste Management:

➢ Establish a system for the proper disposal of electronic waste, including batteries, old components, and chemicals.

Chapter 3: Basic Electronics Theory

Chapter 3 is a critical component of electronics repair at home. It lays the groundwork for understanding how electronics work, allowing you to make informed diagnoses and repairs.

1. Introduction to Electronics Principles

Electronics is a fascinating field that powers many aspects of our modern lives, from the smartphones we carry in our pockets to the computers we use daily and the appliances that make our homes more comfortable. At its core, electronics is built upon a few fundamental principles that serve as the foundation for understanding how electronic circuits and devices function. In this introduction, we'll explore three crucial concepts: voltage, current, and resistance, and their roles in the world of electronics.

- **Voltage - The Driving Force**

Voltage, often denoted as 'V' and measured in volts (V), is the driving force behind electronic circuits. Think of voltage as electrical pressure, similar to water pressure in a hose. It represents the potential energy that pushes electrical charge through a circuit. Just as higher water pressure forces water to flow through a hose, higher voltage encourages electric charge to move through a conductor.

Voltage is what makes your electronic devices come to life. When you press the power button on your TV remote or plug your smartphone into a charger, you're creating a voltage difference that motivates electric charge to flow, providing the energy needed to perform various functions.

- **Current - The Flow of Charge**

Current, represented as 'I' and measured in amperes (A), is the actual flow of electric charge in a circuit. It's akin to the flow of water in a hose. Current tells us how many electric charges (usually electrons) are moving past a point in the circuit per unit of time.

Current is what allows electronic devices to perform specific tasks. For instance, in your smartphone, current flows through the circuits, powering the screen, processor, and other components, enabling you to make calls, browse the internet, and play games.

- **Resistance - The Obstacle to Current Flow**

Resistance, denoted as 'R' and measured in ohms (Ω), is the property that opposes the flow of electric charge. It's analogous to the narrowing of a water pipe, which restricts the flow of water. In electronics, resistance hinders the flow of current through a circuit.

Resistance plays a crucial role in regulating the amount of current in a circuit. Components like resistors are intentionally included in electronic devices to control the flow of current and ensure that devices operate safely and efficiently. For example, in a flashlight, the resistor limits the current through the bulb to prevent it from burning out.

In summary, voltage, current, and resistance are the fundamental building blocks of electronics. Voltage provides the driving force, current represents the flow of charge, and resistance acts as an obstacle to that flow. Together, these principles allow us to create intricate electronic circuits and devices that power our modern world. As you delve deeper into the world of electronics, you'll discover how these principles interact to design, build, and repair a wide range of electronic systems.

2. Understanding Circuits:

Electronic circuits are the lifeblood of modern technology, enabling a wide array of devices to function, from your smartphone to your computer and everything in between. These circuits are constructed using a combination of electronic components that work together to control the flow of electricity and perform specific functions. In this explanation, we'll explore the basic structure of electronic circuits, the interconnection of components, and the roles that key components like resistors, capacitors, diodes, and transistors play in these circuits.

- Basic Structure of Electronic Circuits:

Electronic circuits are composed of interconnected electronic components arranged on a circuit board or within an integrated circuit (IC) package. At the heart of these circuits is the printed circuit board (PCB), a flat board made of insulating material with thin layers of conductive copper traces. The components are soldered onto these traces, forming the circuit.

- Interconnection of Components:

The interconnection of components in an electronic circuit is crucial for proper functionality. Components are linked using conductive traces on the PCB, wires, or metal pins in the case of ICs. Components can be connected in series or in parallel, and the specific arrangement determines the circuit's behavior.

a) **Series Connection**: Components are connected one after the other in a single path. In a series circuit, the same current flows through each component, and the total resistance is the sum of individual resistances.

b) **Parallel Connection**: Components are connected across a common node, creating multiple paths for current. In a parallel circuit, the voltage across each component is the same, but the total current is divided among the branches.

- Key Components and Their Functions:

a) **Resistors**: Resistors are components that restrict the flow of current in a circuit. They are commonly used to limit current, divide voltage, and protect components. The resistance value, measured in ohms (Ω), determines the degree of current restriction.

b) **Capacitors**: Capacitors store and release electrical energy in the form of charge. They are used to filter signals, store energy, and provide timing in circuits. The capacitance value, measured in farads (F), specifies the amount of charge a capacitor can store.

c) Diodes: Diodes allow current to flow in one direction while blocking it in the opposite direction. They are used for rectification (converting AC to DC), voltage regulation, and signal protection.

d) Transistors: Transistors are semiconductor devices that amplify or switch electronic signals. They serve as amplifiers in audio systems, switching devices in digital logic circuits, and are the core of many electronic applications.

These components work together to create a variety of circuits that serve different functions. For example, a simple LED circuit might consist of a resistor to limit current and a diode (the LED) to emit light when current flows. In more complex circuits, such as those in your computer, a combination of resistors, capacitors, diodes, and transistors work in tandem to process and transmit data.

Understanding the structure and function of these components and how they interconnect is fundamental to grasping the operation of electronic circuits and to design, repair, or troubleshoot electronics effectively.

3. Ohm's Law and Practical Applications:

Ohm's Law is a fundamental principle in electronics that establishes a direct relationship between voltage (V), current (I), and resistance (R) in an electrical circuit. It's a powerful tool for understanding and troubleshooting electronic circuits. Let's delve into Ohm's Law and explore practical applications of this principle.

Ohm's Law: $V = I \times R$

Voltage (V): Voltage, often measured in volts (V), represents the electrical potential difference between two points in a circuit. It's akin to the pressure in a water pipe. Voltage provides the driving force for electrical current to flow.

Current (I): Current, measured in amperes (A), is the rate of flow of electric charge through a conductor. It's similar to the rate of water flow in a hose. Current is what makes electrical devices operate.

Resistance (R): Resistance, measured in ohms (Ω), is a property that opposes the flow of electrical current. It's akin to the narrowing of a water pipe, which restricts water flow. Resistance is what limits the flow of current in a circuit.

- **Practical Applications of Ohm's Law**:

Calculating Voltage: You can use Ohm's Law to calculate the voltage drop across a resistor or any other component in a circuit. For instance, if you know the current flowing through a resistor and its resistance, you can find the voltage drop using $V = I \times R$.

Determining Current: Ohm's Law can help you determine the current in a circuit. If you know the voltage and resistance, you can calculate the current using $I = V / R$. This is especially useful when you need to determine the current flowing through a component like an LED or a resistor.

Choosing the Right Resistor: When designing or repairing circuits, you may need to select an appropriate resistor to limit current. By rearranging Ohm's Law as $R = V / I$, you can calculate the resistance needed to achieve a specific current flow at a given voltage.

Power Calculations: Ohm's Law can be extended to calculate power (P) in a circuit. The formula is P = V x I. This is crucial for understanding the power requirements and dissipation in electronic components, which can help prevent overheating and damage.

Troubleshooting Circuits: When a circuit isn't behaving as expected, Ohm's Law can assist in diagnosing the problem. By measuring voltage, current, and resistance at various points in the circuit, you can identify where the issue lies. If, for instance, the measured voltage and current don't match the expected values according to Ohm's Law, there might be a problem with a component or connection.

LED Resistor Calculation: Ohm's Law is often used to calculate the appropriate resistor to limit the current through an LED to its specified value, ensuring the LED operates correctly and doesn't get damaged.

In electronics, Ohm's Law is a versatile tool that allows you to make informed decisions about component selection, troubleshoot circuit problems, and understand how voltage, current, and resistance interact in various situations. It's an essential principle for anyone working with electronics, from hobbyists to professionals.

4. Measuring and Testing with a Multimeter:

A multimeter is an essential tool for measuring and testing electrical properties in electronic circuits. It allows you to measure voltage, current, resistance, and continuity, making it a versatile instrument for troubleshooting and working with electronics. Here, we'll discuss how to use a multimeter for these measurements and provide practical examples of measuring values in electronic circuits.

1. Measuring Voltage:

Set the multimeter to the voltage (V) setting: Turn the dial to the voltage setting with the appropriate range. For example, if you expect to measure 5V, set the multimeter to a range slightly higher than 5V.

Connect the multimeter: Insert the black (common) lead into the COM (common) port and the red (positive) lead into the VΩmA (voltage, ohms, milliamps) port.

Measure voltage: Place the black lead on the circuit's ground (usually the negative side) and the red lead at the point where you want to measure the voltage. Read the voltage value displayed on the multimeter screen.

Practical Example: To measure the voltage of a battery, place the red lead on the positive terminal and the black lead on the negative terminal. The multimeter will display the battery's voltage.

2. Measuring Current:

Set the multimeter to the current (A) setting: Turn the dial to the current setting with the appropriate range. For example, if you expect to measure a current of 100mA, set the multimeter to a range slightly higher than 100mA.

Connect the multimeter: Insert the black lead into the COM (common) port and the red lead into the VΩmA (voltage, ohms, milliamps) port.

Interrupt the circuit: To measure current, you must break the circuit and insert the multimeter in series with the load (device or component). Connect the red lead to the point where the current enters the circuit and the black lead to the point where it exits. The multimeter will display the current value.

Practical Example: To measure the current flowing through an LED, disconnect one of the LED's legs and insert the multimeter in series between the LED and its power source.

3. Measuring Resistance:

Set the multimeter to the resistance (Ω) setting: Turn the dial to the resistance setting with an appropriate range. Start with a higher range and decrease it as needed for accuracy.

Connect the multimeter: Insert the black lead into the COM (common) port and the red lead into the VΩmA (voltage, ohms, milliamps) port.

Measure resistance: Place the leads at the two points across the component or section of the circuit for which you want to measure resistance. The multimeter will display the resistance value.

Practical Example: To check the resistance of a resistor, place the leads on each end of the resistor. The multimeter will show its resistance value.

4. Testing Continuity:

Set the multimeter to the continuity or diode testing setting: This setting often includes a diode symbol or a sound icon (indicating an audible continuity test).

Connect the multimeter: Insert the black lead into the COM (common) port and the red lead into the port labeled for continuity or diode testing.

Test continuity: Touch the two leads together. If the circuit or component has continuity (i.e., it's not open or broken), the multimeter will produce an audible beep or display a numerical value close to zero.

Practical Example: To check if a wire or connection is continuous, place one lead at each end of the wire. If the wire is intact, the multimeter will beep or show continuity.

Multimeters are invaluable tools for diagnosing electronic circuit issues, verifying component values, and ensuring proper connections. With practice, they become essential companions for electronics enthusiasts and professionals alike.

Chapter 4: Identifying Common Electronic Components

Chapter 4 focuses on familiarizing you with the essential electronic components you will encounter when repairing electronics. This chapter aims to provide you with the knowledge and skills to identify, assess, and work with common components.

1. Recognizing Electronic Components:

Electronic components are the building blocks of electronic circuits, and understanding their types and functions is essential for anyone working with electronics. Here, we'll introduce you to some of the most frequently used electronic components, including resistors, capacitors, diodes, transistors, and integrated circuits, and provide clear explanations and visual aids to help you recognize them.

a) Resistors:

Appearance: Resistors are typically small cylindrical components with two wire leads extending from each end.

Function: Resistors restrict the flow of electrical current in a circuit. They are used to control current, divide voltage, set biasing levels, and limit current through components like LEDs.

b) Capacitors:

Appearance: Capacitors come in various shapes and sizes, but common ones are cylindrical or rectangular with two wire leads.

Function: Capacitors store and release electrical energy in the form of charge. They are used for filtering signals, storing energy, and providing timing in circuits.

c) Diodes:

Appearance: Diodes are typically small, cylindrical components with one colored band (the cathode) or a flat edge (the cathode) to indicate polarity.

Function: Diodes allow current to flow in one direction while blocking it in the opposite direction. They are used for rectification (converting AC to DC), voltage regulation, and signal protection.

d) Transistors:

Appearance: Transistors are small, flat, three-legged components. They come in various types, such as NPN and PNP bipolar junction transistors (BJTs) and N-channel and P-channel metal-oxide-semiconductor field-effect transistors (MOSFETs).

Function: Transistors amplify or switch electronic signals. They are used as amplifiers in audio systems, switching devices in digital logic circuits, and form the core of many electronic applications.

e) Integrated Circuits (ICs):

Appearance: Integrated circuits come in various packages, but most are small, flat chips with multiple pins for connecting to a circuit board.

Function: Integrated circuits are complete circuits that perform specific functions, such as microprocessors, amplifiers, or memory. They can contain thousands or even millions of transistors and other components.

Recognizing these common electronic components is a fundamental step in working with electronics. These components come in various sizes, values, and specifications, so understanding their functions and how to identify them visually is crucial for designing, repairing, or troubleshooting electronic circuits.

2. Component Markings and Values:

Electronic components often bear markings that provide information about their value, tolerance, and purpose. For resistors and capacitors, these markings are essential for identifying the component's characteristics. Let's explore how to read these markings and understand their significance.

a) Reading Resistor Markings:

Color Code: Resistors often use a color code to indicate their resistance value. This code consists of color bands on the resistor body. Here's how to interpret it:

First Band: The first band represents the first digit of the resistance value.

Second Band: The second band represents the second digit of the resistance value.

Third Band: The third band indicates the multiplier, which multiplies the first two digits to get the resistance value. For example, if the third band is red (100), it multiplies the first two digits by 100.

Fourth Band (optional): The fourth band represents the tolerance, indicating how much the resistor's actual value may deviate from the specified value.

For example, a resistor with the colors brown (1), black (0), red (100), and gold (±5%) has a resistance value of 10 x 100 = 1000 ohms with a tolerance of ±5%.

Alphanumeric Code: Some resistors use an alphanumeric code. In this system, the resistor is marked with alphanumeric characters that represent the resistance value and tolerance.

The most common alphanumeric code is the EIA-96 standard.

b) Reading Capacitor Markings:

Value: Capacitors are often marked with a numerical code that represents their capacitance value. For example, a capacitor marked "104" has a capacitance of 10 x 10^4 picofarads, or 100,000 picofarads (or 0.1 microfarads).

Tolerance: Some capacitors include a tolerance value, usually represented as a letter. Common tolerance values include J (±5%) and K (±10%). For instance, a capacitor marked "104J" has a capacitance of 100,000 picofarads with a tolerance of ±5%.

Voltage Rating: Capacitors also indicate their voltage rating, which represents the maximum voltage they can handle without breaking down. This is typically written as a voltage value followed by "V" or a voltage symbol. For example, "25V" means the capacitor has a 25-volt rating.

Understanding these markings is crucial for selecting the right components for your circuit and for troubleshooting. Incorrectly reading these markings could lead to circuit malfunction or damage. Always refer to datasheets or component specifications when in doubt.

3. Testing and Measuring Components with a Multimeter:

A multimeter is a versatile tool for testing and measuring electronic components. It allows you to assess the functionality and characteristics of various components. Here's how to use a multimeter to measure resistance, capacitance, and diode functionality:

a. Measuring Resistance:

Set your multimeter to the resistance (Ω) mode. If your multimeter has multiple resistance ranges, select one that's higher than the expected resistance of the component you're testing. Start with the highest range and decrease it as needed for accuracy.

For resistors, place the component between the multimeter's test leads.

If you're testing an SMD (Surface Mount Device) resistor or a component in-circuit, make sure the circuit is de-energized to avoid false readings.

Read the resistance value on the multimeter's display. If the component is functional, you'll see a reading close to its labeled resistance value. If the reading is extremely high or close to infinity, the component is likely open (faulty). If the reading is near zero, it's likely shorted (also faulty).

b. Measuring Capacitance:

Set your multimeter to the capacitance (usually denoted by a farad symbol) mode.

Connect the multimeter's test leads to the capacitor. Ensure that you've discharged the capacitor (if applicable) to avoid any unexpected discharges.

The multimeter will display the capacitance value. If it's a polarized capacitor, ensure that you connect the positive and negative leads correctly, as reversed voltage can damage these components.

c. Testing Diode Functionality:

Set your multimeter to the diode testing mode, often denoted by a diode symbol or a sound icon.

Connect the multimeter leads to the diode. If it's a discrete diode, place the red lead on the anode (the side with the band or mark) and the black lead on the cathode (the side without the band or mark). If it's an SMD diode, it doesn't matter which lead you connect to which side.

If the diode is functional, the multimeter will typically display a voltage drop value (around 0.5 to 0.7V) and may beep. The direction of the voltage drop should be from the anode to the cathode.

If the diode is reversed or open (faulty), the multimeter will display "OL" (open loop) or show a high voltage value in one direction and a low value in the other direction.

Remember that multimeters have different functionalities and features, so always consult your multimeter's user manual for specific instructions and to understand how to interpret the readings. When testing components in-circuit, it's important to de-energize the circuit to avoid inaccurate readings and ensure safety.

4. Troubleshooting with Component Knowledge:

Component recognition is indeed crucial for troubleshooting electronic devices. Knowing the components, their functions, and their expected values allows you to diagnose and fix issues effectively. Here's how component recognition plays a pivotal role in troubleshooting and how a component's failure or incorrect value can cause device malfunctions.

a. Identifying Component Failures:

Resistors: When resistors fail (usually by opening up), they can disrupt the expected flow of current in a circuit. If a current-limiting resistor in an LED circuit fails, the LED may get too much current and burn out. Recognizing this faulty resistor helps in pinpointing the issue.

Capacitors: Failed capacitors can lead to issues like power supply instability or audio distortion. If you know the function of capacitors in a circuit and recognize one that is leaking or swollen, you can replace it to resolve the problem.

Diodes: A failed diode can disrupt the proper direction of current flow, affecting rectification or signal control. If you recognize a diode's function and identify one that's not allowing current in the expected direction, you can replace it.

Transistors: Transistors are often used as switches or amplifiers. If you understand their function and find one that's not amplifying or switching correctly, you can suspect transistor failure.

Integrated Circuits: Integrated circuits serve various functions. If you recognize the type of IC and understand its role in the circuit, you can identify issues related to it. For example, if a microcontroller isn't performing as expected, recognizing it is crucial for troubleshooting.

b. Understanding Expected Component Values:

Resistors: Recognizing resistor values ensures they perform their intended function. If a 1k-ohm resistor is supposed to limit current in an LED circuit and you find a 10-ohm resistor in its place, you'll understand why the LED is not working as expected.

Capacitors: Capacitors store and release energy based on their capacitance values. If you know the expected capacitance and find a capacitor with a significantly different value, you can deduce potential issues with timing or energy storage.

Diodes: Understanding the voltage drop and forward current values for diodes helps you identify if a diode is incorrectly rated or damaged.

Transistors: Recognizing the type of transistor (NPN, PNP, N-channel, P-channel, etc.) and its specifications (gain, voltage rating) ensures you replace it with the correct part when needed.

Integrated Circuits: Knowing the IC's function and specifications is essential for troubleshooting. If a microcontroller is not responding correctly, recognizing its role in the circuit allows you to focus your diagnosis.

c. Replacing Faulty Components:

Once you've recognized a failed or incorrectly valued component, you can replace it with the correct part. This replacement is often the key to fixing electronic devices. It's worth noting that recognizing components and their expected values can save time and resources by preventing unnecessary replacements and focusing your efforts on the likely culprits.

In summary, component recognition is a fundamental skill in troubleshooting electronic devices. Understanding components' functions and their expected values enables you to diagnose and resolve issues efficiently, ensuring that electronic devices operate as intended.

Chapter 5: Troubleshooting Techniques

Chapter 5 delves into the practical aspects of diagnosing and addressing issues in electronic devices. This chapter provides you with valuable skills in troubleshooting.

1. The Process of Troubleshooting:

Troubleshooting is a systematic approach to problem-solving, particularly in the context of identifying and resolving issues in electronic devices or systems. The process involves a series of steps that help diagnose and correct problems efficiently. Here is a structured process for troubleshooting:

1. Observation:

Identify the Problem: The first step is to observe and define the problem. Start by collecting information about what's not functioning correctly or as expected. This may involve talking to users or observing the malfunctioning device.

2. Information Gathering:

Gather Data: Collect relevant information about the problem. This includes any error messages, symptoms, or unusual behavior. If the issue is related to an electronic component or system, make a note of the specific conditions under which the problem occurs.

3. Identification:

Identify Potential Causes: Based on the observations and gathered information, create a list of potential causes or reasons for the problem. This list should include both common issues and less likely but possible causes.

4. Hypothesis Formulation:

Formulate Hypotheses: Develop educated guesses or hypotheses about the possible causes of the problem. Hypotheses should be specific and testable. Each hypothesis should relate to one of the potential causes identified in the previous step.

5. Testing Hypotheses:

Testing and Validation:
Systematically test each hypothesis. This involves performing experiments, measurements, or checks to determine if a particular hypothesis is correct. If the test confirms a hypothesis, it may be the root cause. If not, move on to the next one.

6. Isolation:

Isolate the Problem: If testing a hypothesis reveals that it's incorrect, you can often rule out a specific cause. Continue testing other hypotheses. As you test and eliminate potential causes, the problem's actual source becomes clearer.

**7. Resolution:

Implement Solutions: Once you have identified the root cause of the problem, take steps to resolve it. This may involve repairing or replacing faulty components, reconfiguring settings, or making necessary adjustments.

**8. Testing After Resolution:

Verify the Solution: After implementing a solution, thoroughly test the system or device to ensure that the problem is resolved. Check if the symptoms have disappeared and that the device is functioning correctly.

**9. Documentation:

Document the Troubleshooting Process: It's crucial to maintain a record of the troubleshooting process, including the observed symptoms, potential causes, hypotheses, tests conducted, and the final resolution. Documentation is valuable for future reference and for sharing insights with others.

**10. Preventive Measures:

Prevent Future Occurrences: If possible, identify preventive measures to avoid similar issues in the future. This might involve adding protective components, changing operating conditions, or providing user guidance.

**11. Feedback and Continuous Improvement:

Seek Feedback: Gather feedback from end-users or colleagues involved in the troubleshooting process. This feedback can offer insights into the effectiveness of the resolution and the overall troubleshooting process.

**12. Continual Learning:

Troubleshooting often provides valuable learning experiences. Continually build your troubleshooting skills by analyzing the outcomes of previous issues and updating your knowledge.

The troubleshooting process is iterative and may require revisiting certain steps if the initial hypothesis testing doesn't identify the problem's root cause. It's important to approach each issue with patience and a logical, methodical mindset to efficiently diagnose and resolve problems.

2. Using Schematics and Service Manuals:

Schematics and service manuals are invaluable resources for troubleshooting and repairing electronic devices. They provide detailed information about the device's components, connections, and operation. Here's why they're important and how to use them:

**1. Importance of Schematics and Service Manuals:

Understanding the Device: Schematics and service manuals help you understand the device's internal structure, components, and how they interact. This knowledge is essential for effective troubleshooting.

Troubleshooting: They provide a visual representation of the device's circuitry, making it easier to locate faulty components and diagnose issues.

Safety: Schematics often include safety information, such as voltage ratings and precautions, which is crucial when working with electronic devices.

2. Reading and Interpreting Schematics:

Components: Schematics use standardized symbols for components like resistors, capacitors, transistors, and ICs. Learn these symbols to understand the schematic.

Connections: Lines or wires on the schematic represent electrical connections. They indicate how components are linked in the circuit.

Power Supply: Look for the power supply source (usually labeled Vcc or Vdd) and ground (often labeled GND or 0V).

Signal Flow: Schematics show how signals or currents flow through the circuit, helping you trace problems in signal paths.

Signal Names: Signal names or labels help you identify where a particular signal originates and where it's headed.

Component Values: Component values like resistor values or capacitor capacitance may be indicated on the schematic, allowing you to cross-reference these values with the actual components in the device.

References: Refer to the component list or bill of materials (BOM) to cross-reference component designators on the schematic with their real-world counterparts.

3. Finding Service Manuals:

Manufacturer Websites: Many manufacturers provide service manuals for their devices on their official websites. Look for a "Support" or "Downloads" section.

Third-Party Websites: Some websites aggregate service manuals for various devices. Be cautious and ensure the source is reputable.

Online Communities: Forums, discussion boards, or social media groups focused on electronics and specific devices may share service manuals and insights.

Libraries: Some larger libraries have technical sections with service manuals for various devices, especially older ones.

Service Centers: Authorized service centers often have service manuals for the devices they repair. They may provide access or copies upon request.

Purchase: In some cases, you may need to purchase service manuals, especially for specialized or less common devices. Check with technical bookstores or online retailers.

Copyright Considerations: Be aware of copyright restrictions when using service manuals, especially if you plan to share them or use them for commercial purposes.

**4. Using Service Manuals:

Table of Contents: Start by reviewing the table of contents to get an overview of what's covered in the manual.

Search for Specific Issues: If you're troubleshooting a specific problem, use the index or search function to find relevant sections in the manual.

Follow Diagnostic Procedures: Many service manuals include step-by-step diagnostic procedures to help identify and solve common issues.

Safety First: Pay attention to safety instructions, especially if the device operates at high voltages. Always take proper precautions when working with electronic devices.

Schematics and service manuals are indispensable tools for anyone involved in electronics repair or troubleshooting. They provide the insights and guidance needed to understand, diagnose, and fix electronic devices effectively and safely.

3. Isolating and Identifying Faults in Electronic Devices:

Isolating issues in electronic devices is a critical part of troubleshooting. The process involves systematically identifying and isolating faults to pinpoint the root cause of a problem. Here are some strategies for isolating issues:

1. Visual Inspection:

Examine the Device: Start with a thorough visual inspection of the device. Look for loose connections, damaged components, burnt areas, or any obvious physical damage.

2. Functional Testing:

Functional Check: Test the device to see if it exhibits any issues or malfunctions. This might include power-on self-tests, boot-up diagnostics, or running specific tasks to reproduce the problem.

Signal Tracing: Follow the path of signals through the device. Use an oscilloscope or logic analyzer to trace signals and identify where they become abnormal or disappear.

3. Logic Analysis:

Logic Probing: Use a logic analyzer or digital oscilloscope to analyze digital circuits. This helps you assess the state of digital signals, detect logic errors, and identify timing issues.

4. Substitution Testing:

Component Substitution: Swap components with known good ones to see if the problem persists. This helps identify faulty components. It's commonly used for components like resistors, capacitors, and transistors.

Device Substitution: If possible, replace the entire device with a known working one to check if the issue is device-specific or a systemic problem.

**5. Isolation Testing:

Isolation of Subsystems: If the device has subsystems (e.g., power supply, control, input/output), isolate and test them one by one to determine which one is causing the issue.

**6. Software Analysis:

Software Debugging: For devices with embedded software, debugging tools and techniques can help identify software-related issues. This may involve debugging code, analyzing logs, or monitoring system variables.

7. Environmental Factors:

Temperature Testing: Test the device at various temperature levels. Some issues may only manifest at certain temperatures.

Humidity Testing: Variations in humidity can also affect electronic components. Testing under different humidity conditions can help identify intermittent problems.

8. Documentation and Records:

Review Service Manuals: Refer to service manuals, schematics, and technical documentation. These resources often provide diagnostic procedures and typical fault scenarios.

Review User Reports: If the device belongs to a user or client, gather information on their observations and experiences. This can provide valuable clues about the problem.

**9. Team Collaboration:

Consult with Colleagues: If you're working in a team or within a community, discussing the issue with others can provide fresh perspectives and potential solutions.

**10. Iterative Process:

- **Systematic Approach**: Troubleshooting often involves an iterative process. After each testing or diagnostic step, evaluate the results and decide the next logical step based on the findings. Continue this process until the problem is isolated.

11. Documentation:

- **Record Findings**: Maintain detailed records of your observations, tests, and findings. Documenting your progress helps in tracking the troubleshooting process and can serve as a reference for future issues.

Effective fault isolation is a combination of systematic testing, thorough documentation, and a good understanding of the device's operation. By following these strategies, you can efficiently identify and isolate issues in electronic devices, leading to successful troubleshooting and repairs.

4. Identifying Faulty Components:

Identifying faulty components in electronic devices is a crucial step in troubleshooting and repair. Faulty components can cause various issues, from circuit malfunctions to complete device failure. Here's a guide on how to identify potentially faulty components based on their physical appearance and measurements, and how this identification can lead to effective repairs:

1. Visual Inspection:

Burn Marks: Look for burn marks, discoloration, or physical damage on components. Burn marks often indicate overheating or a short circuit.

Swelling or Leaking: Capacitors, especially electrolytic ones, can swell or leak when they fail. Swollen or leaky capacitors should be replaced.

Cracks or Physical Damage: Cracked or physically damaged components are likely faulty. This includes cracked resistors, damaged diodes, or chipped ICs.

Loose Connections: Check for loose or poorly soldered connections. A loose connection can lead to intermittent faults.

**2. Smell and Sound:

Burning Odor: If you detect a burning or unusual odor when the device is powered on, it may indicate a component is overheating or failing.

Popping or Crackling Sounds: Unusual sounds during operation can also suggest a component failure, such as a capacitor discharge or a transistor malfunction.

**3. Measurement Techniques:

Multimeter: Use a multimeter to measure resistance, capacitance, and voltage across components. If you get readings significantly different from their expected values or no readings at all, the component may be faulty.

Oscilloscope: An oscilloscope can help you analyze waveforms and voltage signals. Abnormal waveforms, noise, or missing signals can point to component issues.

Logic Analyzer: For digital circuits, a logic analyzer can help analyze and decode digital signals. Illogical or inconsistent signal patterns can indicate problems.

Thermal Imaging: Infrared thermography can detect hotspots on components, revealing overheating components. These hotspots may indicate faulty components.

**4. Replacement and Testing:

Component Substitution: One effective way to identify faulty components is to replace them with known good ones. If the problem disappears after replacement, the original component was likely faulty.

Continuity Testing: Use a multimeter to test for continuity in traces or wires. A lack of continuity can indicate a broken or damaged component.

**5. Diode Testing:

Diode Test Mode: Most multimeters have a diode testing mode. Test diodes to check if they conduct in one direction and block in the other. If a diode fails this test, it's likely faulty.

**6. Capacitance Testing:

Capacitance Measurement: Use a multimeter's capacitance mode to measure capacitance values of capacitors. If the measured value is significantly different from the labeled value, the capacitor may be faulty.

**7. Resistance Measurement:

Resistor Values: Measure resistance values of resistors with a multimeter. If the measured value is significantly different from the labeled value, it's likely faulty.

**8. Voltage Measurement:

Voltage Checks: Monitor voltage levels at various points in the circuit using a multimeter or oscilloscope. Abnormal voltage levels can indicate problems in specific areas.

Identifying faulty components through these methods is a key part of the troubleshooting process. By recognizing and replacing or repairing these components, you can often resolve the underlying issues in electronic devices, leading to effective repairs and proper functioning.

Chapter 6: Soldering and Desoldering

Chapter 6 focuses on the practical skills of soldering and desoldering, essential techniques for repairing and modifying electronic components. This chapter is designed to equip you with the knowledge and skills needed to handle electronic connections effectively.

1. The Art of Soldering:

Soldering is a fundamental skill in the world of electronics and is the process of joining two or more metal components using a filler material, called solder, to create a strong, electrically conductive connection. This technique is vital for building and repairing electronic circuits and devices. Let's explore the importance of soldered connections in electronic circuits and devices:

**1. Creating Reliable Electrical Connections:

Soldering creates reliable electrical connections that conduct current efficiently. These connections are essential for the proper functioning of electronic circuits. Solder joints provide a low-resistance path for electrical signals, preventing signal loss and ensuring that components work as intended.

**2. Mechanical Strength:

Soldered connections provide mechanical strength to hold components in place. This is particularly important in printed circuit boards (PCBs), where components are soldered to the board's surface. Solder forms a durable bond, preventing components from loosening due to vibrations or thermal expansion and contraction.

3. Space Efficiency:

Soldered connections are compact and space-efficient. In modern electronics, miniaturization is critical, and soldering allows for the creation of compact, densely populated circuit boards. This space-saving feature is essential in applications where size and weight constraints are a concern, such as mobile devices.

4. Conductivity:

Soldered joints offer excellent electrical conductivity. They conduct electricity with minimal resistance, ensuring that electronic signals flow smoothly between components. This is crucial for maintaining signal integrity and minimizing power loss.

5. Heat Dissipation:

Soldered connections efficiently transfer heat. In electronic circuits, some components generate heat, such as transistors or integrated circuits. Soldered connections allow heat to flow away from these components and dissipate into the PCB or other heat sinks, preventing overheating.

6. Durability:

Soldered connections are durable and resistant to environmental factors like moisture and oxidation. This durability ensures the long-term functionality of electronic devices, even in challenging conditions.

7. Repairability:

Soldered connections are repairable. When a component fails or needs replacement, desoldering and resoldering are common repair processes. Proper soldering techniques allow for the easy removal and replacement of components.

8. Customization:

Soldering allows for the customization of electronic circuits. Components can be precisely positioned and interconnected as required, providing flexibility for designing and building custom electronics.

9. Quality Control:

Soldering is a well-controlled and repeatable process. This consistency is important in manufacturing electronic devices where quality and reliability are paramount.

10. Historical Significance:

Soldering has played a pivotal role in the evolution of electronics. It has been used in electronics manufacturing for over a century and continues to be a fundamental technique in creating a wide range of electronic devices.

In summary, soldering is a critical skill for anyone involved in electronics, from hobbyists to professionals. It plays a central role in creating reliable, compact, and durable electronic circuits and devices, ensuring they operate as intended and stand up to the demands of various applications. Whether you're building a DIY project or repairing a smartphone, mastering the art of soldering is essential for success in the world of electronics.

2. Selecting the Right Soldering Iron:

Choosing the appropriate soldering iron is crucial for achieving successful soldering results. The right soldering iron will depend on the specific tasks you need to perform. Here are some key considerations when selecting a soldering iron:

**1. Wattage:

Low-Wattage (15-40W): These are suitable for delicate electronic components and PCB work. They heat up quickly and are less likely to damage sensitive components. However, they may not provide enough heat for larger or heat-absorbing components.

Medium-Wattage (40-80W): These are versatile for general electronic work and are suitable for most through-hole soldering tasks. They offer a good balance between precision and power.

High-Wattage (80W and above): These are used for heavy-duty applications, such as soldering thick wires, ground planes, or large connectors. They provide ample heat for quick soldering but can be overkill for most electronic work.

2. Temperature Control:

Fixed Temperature: Some soldering irons operate at a fixed temperature. These are suitable for basic soldering tasks but may not be ideal for tasks that require precise temperature control.

Adjustable Temperature: Soldering irons with adjustable temperature settings offer more flexibility. They are essential for working with a variety of components and solder types. Look for a soldering iron with a clear temperature scale and easy adjustment controls.

**3. Tip Types:

Chisel Tips: Chisel tips are versatile and provide a larger flat surface, making them suitable for a wide range of soldering tasks, including through-hole components and SMD work.

Conical Tips: Conical tips are pointed and are ideal for precision work and reaching tight spots on a circuit board.

Bevel Tips: Bevel tips combine the advantages of chisel and conical tips, offering a fine point for precision work and a flat edge for broader coverage.

Specialty Tips: Some tasks may require specialty tips, such as hooved tips for drag soldering or blade tips for soldering surface mount components. Choose a soldering iron that allows for easy tip replacement.

**4. Temperature Recovery:

A soldering iron with good temperature recovery is important for maintaining a consistent temperature during soldering. Quick heat recovery ensures that the iron remains at the desired temperature, even when soldering large or heat-absorbing components.

**5. Soldering Station vs. Standalone Iron:

A soldering station includes a base unit that holds the iron and provides temperature control. Soldering stations are more precise and offer better temperature stability. Standalone irons, on the other hand, are more portable and may not provide as fine-tuned control.

6. Ergonomics:

Consider the ergonomics and comfort of the soldering iron's handle. A comfortable handle design reduces hand fatigue during long soldering sessions.

7. Safety Features:

Look for safety features like auto-shutoff or sleep mode to prevent overheating and potential fire hazards. Additionally, a heat-resistant stand or holder is important for safely placing the iron when not in use.

**8. Quality and Brand:

Invest in a quality soldering iron from a reputable brand. While it may cost more initially, a high-quality soldering iron will last longer, perform more reliably, and provide better soldering results.

When selecting a soldering iron, it's important to match the iron's specifications to the specific tasks you plan to undertake. By considering factors such as wattage, temperature control, tip types, and safety features, you can choose the right soldering iron for your needs and ensure successful soldering projects.

3. Solder Types and Flux:

Solder and flux are integral components in the soldering process, and choosing the right types for specific applications is crucial. Let's explore the different types of solder and flux and their suitability for various tasks:

- ## **Solder Types:**

Lead-Based Solder (Pb-Sn):

Composition: Contains a mixture of lead (Pb) and tin (Sn).

Suitability: Lead-based solder has been widely used for decades. It offers good flow properties and is suitable for general electronic soldering tasks.

Note: Due to environmental and health concerns, many regions have restricted or banned the use of lead-based solder in consumer electronics.

Lead-Free Solder:

Composition: Lead-free solder alternatives often contain tin, silver, copper, or other elements.

Suitability: Lead-free solder is preferred for environmental and health reasons. It is suitable for electronic applications, and specific compositions are designed for various tasks. However, it may require higher soldering temperatures and may be less forgiving in terms of workability.

Flux-Core Solder:

Composition: Solder wire with a flux core.

Suitability: Flux-core solder is convenient for most soldering tasks, as it combines solder and flux in one product. It's suitable for general through-hole and SMD soldering.

Rosin-Core Solder:

Composition: Solder wire with a rosin (resin) core.

Suitability: Rosin-core solder is commonly used in electronics. It has flux embedded in the core, providing flux during soldering. Different flux formulations are available for various applications, such as RMA (Rosin Mildly Activated), RA (Rosin Activated), and No-Clean.

Water-Soluble Flux Solder:

Composition: Contains a water-soluble flux core.

Suitability: Water-soluble flux is easy to clean with water, making it suitable for applications where post-solder cleaning is important, such as in PCB assembly.

No-Clean Flux Solder:

Composition: Contains a no-clean flux core.

Suitability: No-clean flux is designed to leave minimal residue after soldering, eliminating the need for extensive post-solder cleaning. It's ideal for applications where cleaning is impractical.

Selecting the Right Solder and Flux:

Choosing the right solder and flux depends on the specific task and environmental factors. Here's how to select the appropriate combination:

Lead-Free vs. Lead-Based: Consider the environmental regulations in your region. If lead-based solder is permitted, evaluate whether lead-free solder is necessary based on your project's requirements.

Flux Type: Choose the appropriate flux type based on the soldering task:

Rosin-Core Flux: Suitable for most electronics work.

Water-Soluble Flux: When thorough cleaning is required.

No-Clean Flux: When minimal post-solder cleaning is acceptable.

Flux Activity: Fluxes come in different levels of activity, such as RMA (mildly activated) and RA (activated). The choice depends on the required soldering cleanliness and the need for flux residue removal.

Solder Wire Diameter: Consider the diameter of the solder wire. Thinner wires (e.g., 0.5mm) are suitable for precise SMD soldering, while thicker wires (e.g., 1mm) are better for larger components and through-hole soldering.

Temperature and Melting Point: Ensure that the solder's melting point matches the soldering iron's temperature range. Lead-free solders often require higher temperatures than lead-based ones.

Application: Choose the solder and flux based on the specific application, whether it's for electronics, plumbing, jewelry, or other tasks.

Manufacturer Recommendations: Follow the manufacturer's recommendations for solder and flux selection to ensure compatibility with your soldering equipment and components.

In summary, the choice of solder and flux is essential for achieving successful soldering results. Consider the type of solder, flux core, and flux activity, as well as any environmental or health regulations, to select the right combination for your specific soldering tasks.

4. Proper Soldering Techniques for Electronic Components:

Soldering electronic components is a fundamental skill for anyone working with electronics. Creating strong, reliable, and clean solder joints is essential for circuit integrity. Here's a step-by-step guide to proper soldering techniques, including tips for lead-free soldering:

- **Tools and Materials:**

Soldering iron with an appropriate tip.

Solder wire (lead-free or lead-based).

Flux-core solder is suitable for most applications.

Soldering stand or holder.

Heat-resistant surface or mat.

Safety glasses (for eye protection).

- **Procedure:**

Step 1: Prepare Your Work Area

Ensure your work area is clean, well-ventilated, and free from flammable materials. Use a heat-resistant surface or mat to protect your workspace.

Step 2: Select the Right Soldering Iron

Choose a soldering iron with the appropriate wattage for your task. For general electronics work, a 25-40W soldering iron is suitable. If you're using lead-free solder, you may need a slightly higher wattage iron due to the higher melting point.

Step 3: Prepare Your Soldering Iron

Plug in your soldering iron and allow it to heat up to the correct temperature for your solder. Lead-based solder typically melts around 180-190°C (360-370°F), while lead-free solder requires a higher temperature, around 220-240°C (430-465°F).

Step 4: Clean and Tin the Soldering Iron Tip

Before starting, clean the soldering iron tip with a damp sponge or brass wire cleaner. Once clean, apply a small amount of solder to the tip, coating it evenly. This process is called "tinning" and helps to transfer heat efficiently.

Step 5: Apply Flux

If your solder is not flux-core or if additional flux is required, apply a small amount of flux to the area where you intend to make the solder joint. Flux improves solder flow and helps reduce oxidation.

Step 6: Heat the Joint

Touch the soldering iron tip to the component lead or pad and the wire or component you are soldering. Heat both the joint's components and the soldering iron tip simultaneously for a few seconds.

Step 7: Feed the Solder

When the joint is sufficiently heated, touch the solder wire to the joint, not the soldering iron. The heat will melt the solder, and it should flow smoothly around the joint, forming a concave or "volcano" shape. Ensure that the solder wets both the component lead and the pad or wire.

Step 8: Remove the Solder

After a second or two, remove the solder wire. Do not blow on the joint or move it while the solder is cooling. Keep both components steady and stationary.

Step 9: Remove the Soldering Iron

Carefully remove the soldering iron. Do not jerk it away, as this may cause a "cold solder joint." Instead, lift the iron gently while ensuring the joint remains undisturbed.

<u>Step 10: Allow Cooling</u>

Allow the joint to cool naturally without any movement. This ensures a strong, reliable solder joint.

- **Tips for Creating Strong and Reliable Solder Joints:**

Ensure the components are clean and free from oxidation or contamination.

Use the right solder type for your project, either lead-free or lead-based.

Avoid excessive solder. A little solder goes a long way; too much solder can lead to bridging or short circuits.

Always heat the joint first and then apply solder directly to the joint. Never heat the solder directly with the iron.

Keep the soldering iron tip clean and well-tinned to transfer heat effectively.

Practice soldering on scrap components or a spare PCB to improve your technique.

By following these proper soldering techniques and tips, you can create strong, reliable solder joints that are essential for maintaining the integrity of electronic circuits. Whether you're using lead-free or lead-based solder, these principles apply to achieving clean and durable solder connections.

5. Desoldering and Component Removal Techniques:

Desoldering is the process of removing solder from electronic components and circuit boards. It's a crucial skill for repairing and salvaging electronic devices. Several methods can be used for desoldering, including desoldering wick, solder suckers (also known as solder pumps or desoldering pumps), and hot air rework stations. Let's discuss these techniques:

a. Desoldering Wick:

What it is: Desoldering wick, also called solder wick or desoldering braid, is a copper braid coated with flux. It comes in various widths and lengths.

How it works: To use desoldering wick, place it on the solder joint you want to remove and heat it with a soldering iron. The flux in the wick helps to lift the molten solder and absorb it into the braid.

Procedure:

Place the desoldering wick over the solder joint you want to remove.

Heat the wick and the solder joint simultaneously with a soldering iron. The solder will melt and be drawn into the wick due to capillary action.

Keep the wick in place until you've removed enough solder from the joint.

Tips: Use a desoldering wick with flux for better results. Ensure the soldering iron is clean and properly tinned.

b. Solder Suckers (Desoldering Pumps):

What they are: Solder suckers or desoldering pumps are hand-operated tools with a spring-loaded piston and a nozzle. They come in various designs, including bulb-type and plunger-type.

How they work: These tools work by creating a vacuum to suck up molten solder from the joint.

Procedure:

Place the nozzle of the solder sucker over the solder joint.

Heat the joint with a soldering iron until the solder is molten.

Quickly release the spring-loaded piston on the solder sucker to create a vacuum, sucking up the molten solder.

Tips: Maintain a firm and stable grip on the solder sucker. Ensure the nozzle completely covers the solder joint.

c. Hot Air Rework Stations:

What they are: Hot air rework stations use a combination of hot air and precise temperature control to remove and reflow solder on components. They are versatile and suitable for SMD (surface mount device) components and complex circuit boards.

How they work: These stations use heated air to melt the solder. The temperature and airflow can be precisely controlled.

Procedure:

Set the hot air rework station to the appropriate temperature and airflow settings.

Direct the hot air onto the solder joint, heating it until the solder is molten.

Use tweezers or a vacuum pen to lift or gently push the component off the board.

Tips: Hot air rework stations are more suitable for SMD components and complex, multi-layer PCBs. Be cautious to avoid damaging nearby components.

General Desoldering Tips:

Use Flux: Adding additional flux to the solder joint can improve desoldering by reducing oxidation and making the solder flow more smoothly.

Patience: Take your time, especially with complex or densely populated PCBs. Rushing can lead to damage.

Clean Equipment: Ensure your desoldering tools, particularly soldering irons and tips, are clean and properly maintained.

Desoldering is an essential skill for electronics repair and rework. The choice of desoldering method depends on the component, circuit board, and personal preference. With practice and the right tools, you can effectively remove components without damaging the board or surrounding parts.

6. Preventing Overheating and Damage During Soldering and Desoldering:

Overheating is a common concern when soldering and desoldering, as it can lead to component and circuit board damage. To achieve successful soldering and desoldering while preventing overheating, follow these important guidelines:

A . Use the Right Soldering Iron and Tip:

Choose a soldering iron with the appropriate wattage for your task. A lower wattage iron is generally safer for delicate components. Ensure the tip size and shape are suitable for the specific soldering or desoldering job.

B . **Control Soldering Iron Temperature:**

Set the soldering iron to the correct temperature for the solder and components you are working with. Using a soldering iron that is too hot can damage components and circuit boards.

C . **Limit Heating Time:**

Minimize the time components are exposed to heat. Soldering and desoldering should be swift processes. Prolonged exposure to heat can damage sensitive components or adjacent parts.

D . **Use Heat Sinks:**

Heat sinks, such as alligator clips or heat-absorbing putty, can be attached to components to dissipate heat away from sensitive areas. They are particularly useful when soldering or desoldering near heat-sensitive components like semiconductors.

E . Apply Flux:

Flux helps solder flow smoothly and improves heat transfer. Use a suitable flux to aid in the soldering process, reducing the time components are exposed to heat.

F . Preheat SMD Components:

With surface mount devices (SMDs), preheating the entire PCB using a hot air rework station can reduce the risk of overheating individual components during rework.

G . Work in a Controlled Environment:

Maintain a clean and well-ventilated workspace. Good ventilation is essential for removing fumes and preventing the inhalation of potentially harmful substances.

H . Minimize Thermal Stress:

Components and circuit boards can be sensitive to thermal stress. Avoid rapid temperature changes, such as placing a hot PCB on a cold surface, which can lead to cracks or delamination.

I . Practice Good Technique:

Develop proper soldering and desoldering techniques to minimize the time the soldering iron is in contact with components. Use a steady hand to avoid unnecessary movement.

J . Inspect for Damage:

After soldering or desoldering, inspect components and circuit boards for signs of damage. Look for discoloration, cracks, or lifted traces.

K . Use ESD Precautions:

Employ electrostatic discharge (ESD) precautions when working with sensitive electronic components. ESD damage can be invisible but still affect the functionality of components.

L . Be Cautious with Chip Components:

Integrated circuits and chip components can be especially sensitive to heat. Use specialized equipment, such as hot air rework stations, when desoldering and reworking these components.

M . Protect Adjacent Components:

When desoldering, use heat shields or heat-resistant tape to protect nearby components from excessive heat exposure.

N . Learn from Experience:

Experience and practice are essential for improving your soldering and desoldering skills. Start with less critical projects to gain confidence before working on more valuable or complex electronics.

By following these precautions and guidelines, you can prevent overheating and minimize the risk of damage during soldering and desoldering processes, ensuring the reliability and longevity of your electronic components and circuit boards.

Chapter 7: Repairing Common Electronic Devices

Chapter 7 is the practical heart of this book, where you learn how to repair a variety of common electronic devices. This chapter aims to provide hands-on guidance for fixing various types of electronics you might encounter in a household.

1. Introduction to Repairing Electronic Devices:

Repairing electronic devices is a valuable skill that can save you money, reduce electronic waste, and extend the life of your gadgets. Whether it's fixing a malfunctioning smartphone, a faulty laptop, or a damaged circuit board, understanding the repair process is essential. Here, we'll explore the importance of comprehending the repair process and why a systematic approach is crucial.

Why Understanding the Repair Process is Important:

Economic Benefits: Repairing electronic devices is cost-effective. Instead of immediately replacing a malfunctioning device, you can often fix the problem for a fraction of the cost. This can lead to significant savings over time.

Environmental Impact: Repairing devices contributes to sustainability by reducing electronic waste. Electronics contain valuable materials and components that can be reused when properly repaired, helping to minimize the impact of electronic waste on the environment.

Skill Development: Learning to repair electronics is a valuable skill that can be applied to various aspects of your life. It not only enhances your problem-solving abilities but also opens up opportunities for DIY projects and even a potential career in electronics repair.

Device Longevity: Extending the life of your electronic devices through repairs ensures that you get the most value from your investments. This is particularly important for expensive devices like laptops, smartphones, and gaming consoles.

Self-Empowerment: Knowing how to repair your electronics empowers you to take control of your devices. You can diagnose issues, make informed decisions about repairs, and be less dependent on costly repair services.

The Systematic Approach to Repair:

Repairing electronic devices is not a haphazard process; it requires a systematic approach to maximize your chances of success. Here's why a structured method is crucial:

Diagnosis: Identifying the problem accurately is the first step. A systematic approach involves thorough testing and observation to pinpoint the issue.

Planning: Once the problem is understood, plan the repair process. Consider the tools, materials, and replacement components you'll need.

Safety: Electronic devices often contain delicate components and pose potential electrical hazards. A systematic approach emphasizes safety precautions to protect both you and the device.

Documentation: Maintain a record of your repair process. This can include photos, notes, or a digital log. Documentation is essential for tracking progress and learning from your experiences.

Step-by-Step Repair: Follow a structured, step-by-step procedure for the actual repair. This minimizes the risk of overlooking crucial details and keeps the process organized.

Testing and Validation: After the repair, thoroughly test the device to ensure it functions as expected. A systematic approach includes validation to confirm that the problem has been solved.

Post-Repair Documentation: Document the results of your repair, including any changes or adjustments you made. This information can be valuable for future reference or sharing your experiences with others.

Continuous Learning: Electronics and repair techniques evolve. A systematic approach encourages continuous learning and improvement as you gain experience.

Understanding the repair process and adopting a systematic approach are key to successful electronic device repair. It allows you to reap the economic, environmental, and personal benefits of fixing your electronics effectively and efficiently. As we delve deeper into the world of electronics repair, you'll discover essential concepts, techniques, and practices that will equip you with the skills to tackle various repair projects.

2. Repairing Televisions and Monitors: Common Issues and Fixes

Televisions and computer monitors can encounter a range of issues, from display problems to power issues and audio glitches. Understanding how to diagnose and repair these common problems can save you money and extend the life of your devices. Here's a guide to identifying and fixing these issues:

a. Screen Problems:

Dead Pixels: Dead pixels appear as black dots on the screen. They can be caused by manufacturing defects or physical damage. Unfortunately, dead pixels are challenging to repair. In some cases, manufacturers may replace screens under warranty for a certain number of dead pixels.

Stuck Pixels: Stuck pixels are individual pixels that remain a single color while the rest of the screen is active. You can try fixing stuck pixels using software tools that rapidly change the pixel colors or gently massaging the pixel with a soft, non-abrasive cloth.

Flickering or Dim Display: Flickering or dim displays may be due to a loose or damaged video cable or a failing backlight in LCD screens. Ensure that all cables are securely connected. If the backlight is the issue, it may need replacement, which can be a complex repair.

b. Power Issues:

Won't Turn On: If your TV or monitor won't turn on, start by checking the power source. Ensure the power cable is connected properly and the outlet is functional. If there's still no power, the internal power supply or other components may be faulty, requiring professional repair.

Power Cycling: TVs and monitors that cycle on and off can have issues with the power supply, overheating, or internal circuitry. Check for adequate ventilation and ensure the power supply board is functioning correctly.

No Video Signal: If the screen remains black with the power indicator on, it could indicate issues with the internal circuitry or a problem with the source signal. Verify the input source and cable, and try resetting the device.

c. Audio Issues:

No Sound: If you have no sound, check the audio output settings on your device and ensure the volume is up. If the issue persists, it could be a problem with the speakers or the audio processing circuitry. You might need professional repair or external speakers as a workaround.

Audio Distortion or Noise: Audio distortion or noise may result from a poor connection, interference, or faulty components. Check the audio cables and connections. If the issue persists, it might require further troubleshooting or professional repair.

d. Common Troubleshooting Steps:

Factory Reset: Sometimes, issues can be resolved with a factory reset. Refer to the device's manual for instructions on how to perform a factory reset.

Firmware/Software Updates: Ensure your TV or monitor has the latest firmware or software updates installed. These updates can address known issues and improve device performance.

Visual Inspection: Physically inspect the device for loose or damaged cables, connectors, and visible damage. Ensure proper ventilation to prevent overheating issues.

DIY Repairs: For skilled DIYers, you can attempt to replace individual components like capacitors, power supplies, or backlights if you have identified the faulty part. However, these repairs can be complex, so proceed with caution.

Safety Note: When working on electronic devices, always unplug the device and discharge any stored electrical charge to avoid electrical shocks. If you're not confident in your repair abilities, or if the issue is complex, it's best to consult a professional technician or the device's manufacturer for repairs or replacements.

By following these guidelines, you can identify and potentially resolve common issues with televisions and computer monitors. However, always prioritize safety, and if in doubt, seek professional help for complex repairs.

3. Repairing Audio Equipment: Common Issues and Fixes

Audio equipment, including amplifiers, speakers, and headphones, can encounter various issues over time. Knowing how to diagnose and repair these common problems can help you extend the life of your audio gear. Here's a guide to identifying and fixing common audio issues:

a. Distorted Sound:

Amplifiers: Distorted sound from an amplifier can result from a variety of factors, including malfunctioning transistors, capacitors, or improper biasing. Check and replace any damaged components, and ensure that the amplifier is properly biased for your speakers.

Speakers: Distorted sound from speakers may be due to damaged speaker cones, voice coils, or crossovers. Inspect the speakers for visible damage and replace any faulty components. Clean or replace crossovers if necessary.

Headphones: Distorted sound in headphones could be caused by damaged drivers or loose connections. Open the headphones carefully and inspect the drivers. If they are damaged, they may need replacement. Ensure all internal connections are secure.

b. Connectivity Problems:

Amplifiers: Connectivity issues in amplifiers can result from loose or corroded input or output connectors. Clean or replace connectors, and ensure all connections are secure. Check input and output sources for compatibility.

Speakers: Connectivity issues with speakers can stem from loose or damaged speaker wire connections. Ensure all wires are properly connected and replace any damaged cables. Check the speaker connectors as well.

Headphones: Loose or damaged headphone cables can cause connectivity problems. Examine the cables for damage and replace them if necessary. Check the headphone jack for any debris or corrosion.

c. No Sound or Audio Dropout:

Amplifiers: If an amplifier produces no sound, it may have blown fuses, a malfunctioning power supply, or damaged output transistors. Check and replace fuses, repair the power supply, and replace faulty components.

Speakers: No sound from speakers can result from damaged drivers, bad crossovers, or faulty wiring. Inspect and replace damaged components, and ensure all wiring is intact.

Headphones: No sound or audio dropout in headphones may be due to a faulty cable, damaged driver, or loose connections. Examine the cable for damage and replace it if needed. Check for loose or damaged connectors inside the headphones.

d. Component Failures:

Amplifiers: Component failures in amplifiers can include blown capacitors, damaged transistors, or malfunctioning ICs. Identify and replace any damaged components with suitable replacements.

Speakers: Speaker component failures can involve damaged cones, voice coils, or crossover components. Replace any damaged components with compatible replacements.

Headphones: Component failures in headphones could be due to damaged drivers, cracked solder joints, or faulty cables. Inspect the drivers, resolder loose connections, and replace damaged components.

e. Hum or Noise:

Amplifiers: Hum or noise in amplifiers may be caused by ground loops, damaged components, or interference. Isolate and eliminate ground loops, and check and replace damaged components.

Speakers: Hum or noise in speakers can be caused by damaged crossovers or interference. Check and replace damaged components, and ensure proper grounding.

Headphones: Hum or noise in headphones can be due to interference, damaged cables, or connectors. Check for damaged cables, connectors, and interference sources. Replace or shield cables if necessary.

Safety Note: When working on audio equipment, always prioritize safety. Ensure that equipment is unplugged or disconnected from power sources. If you're unsure of your repair skills, or if the issue is complex, consult a professional technician or the device's manufacturer for repairs or replacements.

By following these guidelines, you can identify and potentially resolve common issues in amplifiers, speakers, and headphones. However, always prioritize safety and consider seeking professional help for complex repairs.

4. Repairing Computers and Laptops: Common Issues and Fixes

Computers and laptops are essential devices in our daily lives, but they can experience a range of problems over time. Knowing how to diagnose and repair common issues, whether they're related to hardware or software, can save you money and keep your devices running smoothly. Here's a guide to diagnosing and fixing issues in computers and laptops:

a. Overheating Problems:

- ***Diagnosis***: Overheating is a common issue in laptops and desktop computers. Symptoms include frequent unexpected shutdowns, fan noise, and a hot laptop or computer case.

- ***Repair***:

  ➢ Dust Removal: Overheating is often caused by dust clogging the cooling system. Open the computer or laptop (following proper precautions and safety measures), and use compressed air to clean the fans, heatsinks, and vents.

➢ Thermal Paste Replacement: If the overheating issue persists, you may need to replace the thermal paste on the CPU and GPU. This is a more advanced procedure and should be done carefully to ensure proper heat transfer.

b. Slow Performance:

- *Diagnosis*: Slow performance can result from various factors, including software issues, insufficient RAM, or a fragmented hard drive.

- *Repair*:

- ➢ Software Cleanup: Remove unnecessary startup programs, run disk cleanup, and defragment your hard drive to optimize performance.
- ➢ RAM Upgrade: If your computer or laptop has limited RAM, consider upgrading to improve multitasking and overall speed.
- ➢ Solid-State Drive (SSD) Installation: Replacing a traditional hard drive with an SSD can significantly boost performance due to faster read/write speeds.

c. Screen Issues:

- *Diagnosis*: Screen problems can manifest as flickering, dead pixels, or a completely blank screen.

- *Repair*:

➢ Cable Check: For laptops, screen issues can be due to loose or damaged display cables. Open the laptop (if you're comfortable doing so) and check the cable connections.
➢ Driver Updates: For desktop computers, screen problems might be resolved by updating graphics drivers.
➢ Screen Replacement: In cases of physical screen damage or persistent issues, screen replacement may be necessary.

d. Connectivity Problems:

- **_Diagnosis_**: Connectivity issues can include problems with Wi-Fi, Bluetooth, or USB ports.

- **_Repair_**:

➢ Driver Updates: Ensure that Wi-Fi and Bluetooth drivers are up to date. Uninstall and reinstall the drivers if necessary.
➢ USB Troubleshooting: For USB issues, try connecting devices to different ports to rule out port problems. Update USB drivers if needed.
➢ Router/Modem Check: If you're experiencing Wi-Fi problems, restart your router or modem. Ensure that other devices can connect to the network.

e. Software Troubleshooting:

- ***Diagnosis***: Software issues can lead to system crashes, error messages, or slow performance.

- ***Repair***:

  ➢ Malware and Virus Scans: Run a reliable antivirus or anti-malware program to detect and remove malicious software.
  ➢ System Restore: Use system restore points to revert your computer's software to a previous state if issues have occurred recently.
  ➢ Reinstallation: As a last resort, consider reinstalling the operating system to eliminate persistent software problems.

f. Hardware Failures:

- ***Diagnosis***: Hardware failures may include malfunctioning hard drives, memory modules, or power supplies.

- ***Repair***:

➢ Data Backup: If you suspect a hard drive failure, back up your data immediately. You may need to replace the hard drive and reinstall the operating system.
➢ Memory Testing: Use memory testing tools to check for memory module issues. Faulty modules should be replaced.
➢ Power Supply Replacement: A failing power supply can lead to

erratic computer behavior.
Replacing it is essential.

Safety Note: Prioritize safety when working with computers and laptops. Disconnect them from power sources, use proper grounding, and follow safety guidelines to avoid electrical hazards.

By following these guidelines, you can identify and potentially resolve common issues in computers and laptops. However, if you're unsure or if the issue is complex, consult a professional technician for repairs or replacements.

5. Repairing Smartphones and Tablets: Common Issues and Fixes

Smartphones and tablets have become integral parts of our lives, but they can encounter various problems, from screen damage to battery issues and water damage. Knowing how to diagnose and repair these common issues can save you money and extend the life of your devices. Here's a guide to identifying and fixing issues in smartphones and tablets:

a. Screen Replacement:

- ***Diagnosis***: Cracked or malfunctioning screens are among the most common issues. Symptoms include shattered glass, unresponsive touchscreens, or display anomalies.

- ***Repair***:

➢ Tools and Parts: Gather the necessary tools and a replacement screen. Ensure the screen is compatible with your device.

➢ Replacement Procedure: Open the device carefully, following guides specific to your model. Disconnect the battery, remove the old screen, and install the new one. Be cautious to preserve other components.

b. Battery Issues:

- *Diagnosis*: Battery problems can manifest as fast battery drain, not holding a charge, or the device shutting down unexpectedly.

- *Repair*:

➢ Battery Replacement: If the battery is old or faulty, consider replacing it. This often involves opening the device, so follow proper safety procedures.

➢ Software Optimization: Ensure your device's software is up to date, and manage background processes to reduce battery drain.

c. Water Damage:

- **_Diagnosis_**: Water damage symptoms include a wet device, screen issues, or erratic behavior.

- **_Repair_**:

➢ Immediate Action: If your device gets wet, power it off, remove the battery (if possible), and disassemble it. Dry all components and circuits carefully.

➢ Rice or Silica Gel: Place the device and its components in a container with rice or silica gel packets for several days to absorb moisture.

➢ Professional Help: Severe water damage might require professional repair, especially if corrosion has occurred.

d. Connectivity Problems:

- *Diagnosis*: Connectivity issues can involve problems with Wi-Fi, Bluetooth, or cellular networks.

- *Repair*:

- ➤ Software Updates: Ensure that your device's software, including network drivers, is up to date.
- ➤ Network Reset: Reset network settings to default, which can resolve many connectivity issues.
- ➤ Component Check: If connectivity problems persist, inspect internal components like antennas and connectors for damage.

e. Software Troubleshooting:

- *Diagnosis*: Software issues can lead to system crashes, slow performance, or unresponsive apps.

- *Repair*:

- ➢ Software Updates: Keep your device's software up to date, as manufacturers release updates to address issues.
- ➢ Factory Reset: As a last resort, perform a factory reset to resolve persistent software problems, but remember to back up your data first.

Safety Note: Prioritize safety when working with smartphones and tablets. Disconnect them from power sources, use proper grounding, and follow safety guidelines to avoid electrical hazards. Water-damaged devices should be disconnected from power immediately to prevent short circuits.

By following these guidelines, you can identify and potentially resolve common issues in smartphones and tablets. However, if you're unsure or if the issue is complex, consult a professional technician for repairs or replacements.

6. Repairing Home Appliances: Common Issues and Fixes

Household appliances, such as microwaves, washing machines, and refrigerators, are essential to our daily lives. When these appliances encounter issues, understanding how to diagnose and repair common problems can save you money and keep your household running smoothly. Here's a guide to identifying and fixing issues in common home appliances:

a. Microwave Repair:

- *Diagnosis*: Microwave issues can range from failure to heat, unusual noises, or error codes.

- *Repair*:

➢ Safety Precautions: Always unplug the microwave and discharge the capacitor to avoid electrical shock.
➢ Magnetron Replacement: A common issue is a faulty magnetron. Replacing the magnetron may resolve heating problems.
➢ Door Switch Replacement: If the microwave doesn't start, check and replace the door switches as they are crucial for safety interlocks.

b. Washing Machine Repair:

- *Diagnosis*: Washing machine problems include leaks, excessive vibration, or failure to drain or spin.

- *Repair*:

➢ Safety Precautions: Disconnect power and water supply to the machine before any repairs.
➢ Leaks: Leaks are often caused by damaged hoses, seals, or pumps. Inspect and replace faulty components.
➢ Excessive Vibration: Level the machine and ensure it's properly balanced. Damaged shock absorbers can also cause vibration issues.
➢ Drain/Spin Issues: Clogs or problems with the pump, drain

hose, or motor may lead to
drainage or spin cycle problems.

c. Refrigerator Repair:

- *Diagnosis*: Refrigerator issues can
  involve temperature problems,
  unusual noises, or leaks.

- *Repair*:

➤ Safety Precautions: Disconnect
  power and turn off the water supply
  if your fridge has an ice maker or
  water dispenser.
➤ Temperature Problems: Ensure the
  condenser coils are clean and the
  evaporator fan is working. A faulty
  thermostat or defrost timer may
  require replacement.

- ➢ Noisy Operation: Check for loose or damaged parts like the evaporator or condenser fan, and inspect the compressor for issues.
- ➢ Leak Detection: Leaks can result from a clogged defrost drain or damaged water supply lines. Clear clogs and replace damaged parts.

d. Dishwasher Repair:

- *Diagnosis*: Dishwasher problems can include poor cleaning, failure to drain, or unusual noises.

- *Repair*:

- ➢ Safety Precautions: Turn off power and water to the dishwasher.
- ➢ Poor Cleaning: Check and clean spray arms, filters, and detergent

dispensers. A malfunctioning wash motor or heating element may require replacement.

- ➤ Drainage Issues: Inspect and clean the drain pump, drain hose, and air gap. Replace damaged components as needed.
- ➤ Unusual Noises: Noises may result from foreign objects in the pump or wash arms, or a worn-out motor or pump.

Safety Note: Prioritize safety when working with home appliances. Always unplug the appliance and turn off its water supply before any repair. Be cautious with electrical and water-related components to prevent hazards.

By following these guidelines, you can identify and potentially resolve common issues in household appliances. However, if you're unsure or if the issue is complex, consult a professional technician for repairs or replacements.

7. Repairing Gaming Consoles: Common Issues and Fixes

Gaming consoles are beloved devices for gamers, but like any electronics, they can experience issues. Knowing how to diagnose and repair common problems can help you get back to your gaming quickly and save money. Here's a guide to identifying and fixing issues in gaming consoles:

a. Overheating Problems:

- ***Diagnosis***: Overheating can lead to unexpected shutdowns, the "Red Ring of Death" for Xbox consoles, or the "Yellow Light of Death" for PlayStation consoles.

- ***Repair***:

➤ Internal Cleaning: Overheating is often caused by dust clogging the console's cooling system. Carefully disassemble the console and use compressed air to clean the fans, heatsinks, and vents.

➤ Thermal Paste Replacement: Over time, thermal paste can dry out and lose its effectiveness. Replacing it with fresh thermal paste can improve heat dissipation.

b. Power Problems:

- *Diagnosis*: Power issues can manifest as a console that won't turn on or unexpected shutdowns.

- *Repair*:

➢ Power Cable and Outlet: Ensure that the power cable is securely connected and the outlet is functional.
➢ Power Supply Check: Verify the functionality of the power supply. In some cases, replacing it may be necessary.

c. Optical Drive Issues:

- *Diagnosis*: Problems with the optical drive can result in difficulty reading discs, strange noises, or no response when a disc is inserted.

- ***Repair***:

  ➤ Disc Cleaning: Start by cleaning the game discs. Ensure they are free from scratches and dirt.
  ➤ Laser Lens Cleaning: Dust on the laser lens can lead to read errors. Use a laser lens cleaning kit to clean the lens. If the problem persists, the laser assembly may need replacement.

d. Software Troubleshooting:

- ***Diagnosis***: Software issues can lead to system crashes, error messages, or difficulty starting games.

- *Repair*:

> System Updates: Keep the console's software up to date. Manufacturers often release updates to address issues.
> Game Updates: Ensure that your games are updated to the latest versions. Outdated games can sometimes cause problems.

e. Controller Problems:

- *Diagnosis*: Controller issues can include unresponsive buttons, stick drift, or connectivity problems.

- *Repair*:

- ➤ Button and Stick Cleaning: Disassemble the controller and clean buttons and analog sticks. Dust and debris can affect performance.
- ➤ Stick Replacement: If you experience stick drift (where the in-game character moves without input), you may need to replace the analog stick or the entire controller.

Safety Note: Always unplug the console before performing any repairs, and take appropriate safety precautions to avoid electrical hazards.

By following these guidelines, you can identify and potentially resolve common issues in gaming consoles. However, if you're unsure or if the issue is complex, consult a professional technician for repairs or replacements.

Chapter 8: Preventative Maintenance and Upgrades

Chapter 8 focuses on the concept of proactive care for electronic devices, including maintenance practices to extend the lifespan of these devices and upgrades to enhance their performance. This chapter aims to provide you with practical insights into keeping their electronics in good condition.

1. The Importance of Preventative Maintenance for Electronic Devices

Preventative maintenance is a fundamental practice for maintaining the reliability, longevity, and efficiency of electronic devices. It involves regularly inspecting, cleaning, and repairing components to prevent issues before they occur. Whether it's your smartphone, laptop, gaming console, or household appliances, proactive care can significantly extend the life and functionality of your devices. Here's why preventative maintenance is crucial:

a. Increased Reliability:

Fewer Breakdowns: Regular maintenance helps identify and address potential issues early. This means fewer unexpected breakdowns, reducing frustration and inconvenience.

<u>Consistent Performance</u>: Well-maintained devices operate more consistently, delivering the performance you expect. This is especially critical for professional users and businesses.

b. Prolonged Longevity:

<u>Extended Lifespan</u>: Preventative maintenance ensures that components remain in good working condition. As a result, the overall lifespan of your devices is extended, reducing the need for premature replacements.

<u>Cost Savings</u>: Longevity means you can get more value from your investment. You won't need to replace devices as frequently, saving you money in the long run.

c. **Improved Efficiency**:

Optimal Performance: Maintained devices operate at their peak efficiency. For electronic devices like computers and smartphones, this means faster processing, longer battery life, and smoother operation.

Energy Efficiency: Well-maintained appliances, such as refrigerators or air conditioners, consume less energy. This not only lowers your utility bills but also reduces your carbon footprint.

d. Reduced Repair Costs:

<u>Minor Repairs</u>: Preventative maintenance often involves addressing minor issues, which are generally less costly to fix than major breakdowns.

<u>Preventing Catastrophic Failures</u>: Regular inspections and cleaning can prevent issues that might otherwise lead to catastrophic and expensive failures.

e. Enhanced Safety:

<u>Electrical Safety</u>: Regular maintenance checks for loose connections, frayed wires, or damaged components, reducing the risk of electrical fires and shocks.

<u>Preventing Malfunctions</u>: Preventative maintenance in household appliances

like ovens or microwaves can help avoid dangerous malfunctions that might cause accidents.

f. Personal Convenience:

Consistency: Maintained devices are reliable and work as expected. You can depend on them for both work and personal tasks.

Peace of Mind: Knowing that your devices are well cared for and less likely to fail provides peace of mind, reducing stress and frustration.

g. Sustainability:

Environmental Impact: Extending the life of electronic devices through preventative maintenance reduces the environmental impact of manufacturing and disposing of electronic waste.

Resource Conservation: Repairing and maintaining devices conserves resources and reduces the demand for new products.

In a world where electronic devices are integral to our daily lives, the importance of preventative maintenance cannot be overstated. It not only saves you money but also contributes to a more sustainable and responsible use of technology. By investing a little time and effort in regular maintenance, you can enjoy the benefits of enhanced device performance, longer lifespans, and the peace of mind that comes from knowing your electronics are in good shape.

2. Regular Maintenance Practices for Electronic Devices

Regular maintenance is the key to keeping your electronic devices in optimal working condition. These practices can be applied to a variety of electronic devices, from smartphones to appliances. By following these routine maintenance steps, you can ensure the longevity and reliability of your electronics:

a. Cleaning:

Exterior Cleaning: Wipe down the exterior of the device with a soft, lint-free cloth to remove dust and fingerprints. For screens, use a microfiber cloth or a screen-safe cleaning solution.

Interior Cleaning: Periodically, open your device (if feasible and safe) and clean the internal components. Remove dust and debris using compressed air or a small brush. This is particularly important for computers and gaming consoles to prevent overheating.

b. Dust Removal:

Cooling Systems: Devices with cooling fans, like laptops and gaming consoles, can accumulate dust in their cooling systems. This dust can impede airflow and cause overheating. Use compressed air to blow out the dust from cooling vents.

c. Inspection:

Visual Inspection: Regularly inspect the device for visible wear and tear, loose connections, or damaged components. Look for loose cables, frayed wires, or cracked casings.

Connectivity Check: Test all ports and connections to ensure they are functioning properly. Loose or damaged connectors can lead to connectivity problems.

d. Software Maintenance:

Software Updates: Keep your device's operating system, drivers, and software applications up to date. Manufacturers often release updates to improve stability and security.

Malware Scans: Run regular malware scans on computers and smartphones to detect and remove potential threats that can harm your device's performance.

e. Battery Maintenance:

Calibration: For devices with rechargeable batteries, like smartphones and laptops, perform occasional battery calibrations. This helps the device accurately gauge the battery's capacity.

<u>Avoid Deep Discharges</u>: Try to avoid completely draining the battery, as this can reduce its overall lifespan.

f. Backup and Data Management:

<u>Data Backup</u>: Regularly back up important data on your devices to prevent data loss in case of hardware failure or accidents.

<u>Storage Cleanup</u>: Delete unnecessary files and applications to free up storage space, which can improve device performance.

g. Safe Handling:

<u>Proper Storage</u>: When not in use, store devices in a cool, dry place away from direct sunlight and extreme temperatures.

<u>Protection</u>: Use cases or covers to protect devices from physical damage and keep them safe during transport.

h. Water Damage Prevention:

<u>Waterproofing Measures</u>: Use waterproof cases or covers for devices that may be exposed to water. Keep devices away from liquids to avoid accidental water damage.

By incorporating these regular maintenance practices into your electronic device care routine, you can extend the life and ensure the reliability of your devices. Preventative maintenance not only saves you money and time but also enhances your overall electronic experience. Remember to consult the manufacturer's recommendations and safety guidelines for specific devices, and when in doubt, seek professional assistance for complex maintenance or repairs.

3. Maintenance Schedules for Electronic Devices

Maintenance schedules are structured plans for performing routine care and checks on electronic devices. They ensure that devices continue to function optimally, have an extended lifespan, and experience fewer unexpected issues. Here are guidelines for how often specific maintenance tasks should be performed based on the type of device:

a. Computers (Laptops and Desktops):

<u>Monthly</u>:

Dust Cleaning: Clean the exterior and remove dust from fans and vents.

<u>Quarterly</u>:

Internal Cleaning: Open the computer, clean the internal components, and inspect for loose cables or damaged parts.

Backup: Back up important data.

Malware Scan: Run a malware scan and update your antivirus software.

<u>Annually:</u>

Thermal Paste Replacement: Consider replacing thermal paste to ensure optimal heat dissipation.

Software Updates: Update the operating system and drivers.

b. Smartphones and Tablets:

Monthly:

External Cleaning: Wipe down the exterior with a microfiber cloth.

App Updates: Update apps for security and performance.

Quarterly:

Battery Calibration: Calibrate the battery if your device allows it.

Annually:

Battery Replacement: Monitor the battery's health, and consider replacing it if necessary.

Operating System Update: Update to the latest version of the operating system.

c. Household Appliances (Refrigerators, Microwaves, Washing Machines, etc.):

Monthly:

Exterior Cleaning: Wipe down the exterior to remove dust and spills.

Refrigerator Coils: Vacuum or brush the refrigerator coils to maintain efficiency.

Microwave Cleaning: Clean the interior and exterior of the microwave.

Quarterly:

Dishwasher Filters: Clean and inspect dishwasher filters.

Washing Machine Maintenance: Clean the lint filter and inspect hoses and connections.

Annually:

Refrigerator Door Seals: Check and replace worn-out door seals.

Washing Machine Drum Cleaning: Clean the drum to prevent odors and mold.

d. Gaming Consoles:

Monthly:

External Cleaning: Wipe down the exterior and remove dust.

Quarterly:

Internal Cleaning: Open the console, clean the internal components, and inspect for loose cables or damaged parts.

Software Updates: Update the console's operating system and game software.

Annually:

Controller Maintenance: Clean and inspect controllers, and replace thumbsticks or buttons if necessary.

e. Audio Equipment (Amplifiers, Speakers, Headphones, etc.):

Monthly:

External Cleaning: Wipe down the exterior and remove dust.

Cable and Connection Check: Inspect and secure cables and connections.

Quarterly:

Internal Cleaning: Open the equipment (if possible), clean the internal components, and inspect for loose cables or damaged parts.

Software/Firmware Updates: Check for updates to amplifiers or digital audio devices.

Annually:

Speaker Drivers: Inspect and clean speaker drivers and replace them if necessary.

Amplifier Maintenance: Clean and maintain amplifiers, including replacing tubes or other components if required.

f. Televisions and Monitors:

Monthly:

Screen Cleaning: Wipe down the screen with a microfiber cloth.

Quarterly:

Cable and Connection Check: Inspect and secure cables and connections.

Software/Firmware Updates: Check for updates to smart TVs or monitors.

Annually:

Dust Removal: Open the device (if possible) and clean the internal components, particularly in the case of projectors.

Remember that these are general guidelines, and specific devices may have unique maintenance requirements. Always consult the manufacturer's recommendations and safety guidelines. When in doubt or when maintenance tasks require advanced knowledge or safety precautions, seek professional assistance.

4. Safety Considerations During Maintenance Procedures

Ensuring safety during maintenance procedures is paramount to prevent accidents and protect both you and your electronic devices. Here are essential safety practices to keep in mind when conducting maintenance:

a. Disconnect Power Sources:

<u>Unplug</u>: Always unplug the device from the electrical outlet or disconnect its power source before starting any maintenance work. This prevents electrical shocks and injuries.

<u>Turn Off and Wait</u>: In the case of devices with built-in batteries, ensure they are turned off and give them time to cool down before beginning maintenance. Removing the power source may involve more than just unplugging the device.

b. Protect Against Electrical Hazards:

<u>Wear Insulated Gloves</u>: If you need to work on live electrical components, wear insulated gloves to protect yourself from electric shocks.

<u>Use Proper Tools</u>: When working with electrical components, use insulated tools designed for electrical work.

c. Prevent Electrostatic Discharge (ESD):

<u>ESD Wrist Strap</u>: When working on sensitive electronics like computers or smartphones, wear an ESD wrist strap to prevent static electricity discharge that can damage components.

<u>ESD Work Mat</u>: Place your device on an ESD-safe work mat to further prevent electrostatic discharge.

d. Work in a Clean, Well-Ventilated Area:

<u>Clean Work Environment</u>: Keep your workspace clean and free from clutter to avoid accidents. Dust and debris can also interfere with your work.

<u>Well-Ventilated:</u> Ensure good ventilation, especially when cleaning or

using cleaning agents, to avoid inhaling fumes or chemicals.

e. Follow Manufacturer Guidelines:

Read the Manual: Always consult the manufacturer's guidelines and user manuals for device-specific maintenance instructions and safety precautions.

f. Properly Store Tools and Materials:

Secure Tools: Store your maintenance tools and materials safely, out of reach of children and pets, to prevent accidents.

g. Don't Overreach Your Expertise:

Know Your Limits: If a maintenance task requires expertise beyond your knowledge or involves potential hazards, consider seeking professional assistance. It's safer to have an expert handle complex issues.

h. Maintain Records:

Documentation: Keep records of your maintenance activities, including dates, tasks performed, and any replacements or repairs. This can be helpful for troubleshooting and warranty claims.

i. Be Cautious with Chemicals:

Hazardous Materials: When using cleaning agents or chemicals, follow safety guidelines and be cautious to prevent exposure or accidents.

j. First Aid Preparedness:

<u>First Aid Kit</u>: Keep a basic first aid kit on hand in case of minor injuries. Know how to use it, and be aware of the location of the nearest medical facility.

k. Double-Check Before Powering On:

<u>Inspection</u>: After completing maintenance tasks, double-check that all components and connections are secure before reconnecting power sources and powering on the device.

l. Wear Protective Gear:

<u>Safety Goggles and Gloves</u>: Depending on the task, consider wearing safety goggles and gloves to protect your eyes and hands.

Prioritizing safety during maintenance procedures is crucial to prevent accidents and costly damage to your electronic devices. Always err on the side of caution and consult professional help when in doubt or when dealing with complex repairs. By adhering to these safety practices, you can enjoy a safer and more efficient maintenance experience.

5. Upgrades and Enhancements for Electronic Devices

Upgrades are a powerful concept in the world of electronic devices, offering numerous benefits that can improve performance, extend functionality, and provide a cost-effective alternative to replacing your devices. Here's an introduction to the idea of upgrades and how they can benefit your electronic devices:

a. Improved Performance:

Boost Processing Power: Upgrading components like the CPU, RAM, or graphics card in computers can significantly enhance performance, making tasks faster and more efficient.

Faster Load Times: Upgrading storage devices, such as switching from a traditional hard drive to a solid-state drive (SSD), can result in quicker boot times and reduced application load times.

Enhanced Graphics: For gaming and multimedia tasks, upgrading the graphics card can provide better visuals and smoother gameplay.

b. Extended Functionality:

Added Features: Some upgrades introduce new features to devices, enhancing their capabilities. For instance, upgrading a smartphone camera module can improve photo and video quality.

Connectivity: Upgrades can provide additional connectivity options, such as adding USB-C or Thunderbolt ports to a computer.

c. Cost-Effective Solutions:

Savings Over Replacement: Upgrades are often more cost-effective than buying a new device. They allow you to get more mileage out of your existing device, delaying the need for a costly replacement.

Environmentally Friendly: By upgrading rather than replacing, you reduce electronic waste and contribute to a more sustainable approach to technology.

d. Device Longevity:

Extended Lifespan: Upgrades can extend the lifespan of your electronic devices, ensuring that they remain relevant and functional for longer.

e. Tailored Solutions:

Customization: Upgrades enable you to tailor your device to your specific needs. You can focus on enhancing the aspects that matter most to you.

Compatibility: You can choose upgrades that are compatible with your device, avoiding the need to adapt to a new interface or ecosystem.

f. Easy to Implement:

User-Friendly: Many upgrades can be performed by users themselves, with online tutorials and guides readily available.

Professional Help: For complex upgrades or those requiring specialized knowledge, professionals are available to assist.

Examples of Common Upgrades:

- Computer Upgrades: Upgrading RAM, replacing or adding storage drives, improving the graphics card, or installing a more powerful CPU.

- Smartphone Upgrades: Replacing the battery, upgrading the camera module, or adding more storage.

- Home Appliances: Replacing worn-out seals on refrigerators, upgrading to a more energy-efficient dishwasher, or improving oven thermostat accuracy.

- Gaming Consoles: Upgrading the hard drive for more storage, replacing thermal paste for better cooling, or modding the console for custom features.

Upgrades are an investment that can provide tangible benefits, making your electronic devices more efficient, functional, and enjoyable to use. Before considering a replacement, explore the possibilities of upgrading to get the most out of your existing devices.

6. Practical Examples of Maintenance and Upgrades for Specific Devices:

a. Cleaning a Laptop's Cooling System:

Why: Over time, laptops can accumulate dust in their cooling systems, leading to overheating and reduced performance. Cleaning the cooling system helps maintain optimal temperatures.

Steps:

- ➤ Turn off the laptop and unplug it.
- ➤ Open the laptop: Refer to the user manual for instructions on accessing the internal components.
- ➤ Use compressed air to blow out dust from the cooling fan and vents. Hold the fan blades still to prevent them from spinning while cleaning.
- ➤ Reassemble the laptop and power it on.

b. Upgrading a Computer's RAM:

Why: Upgrading RAM can boost a computer's performance, making it more responsive and capable of handling demanding tasks.

Steps:

> Check compatibility: Ensure the new RAM is compatible with your computer's motherboard.
> Turn off the computer and unplug it.
> Open the computer case: Most desktops have easy-to-access RAM slots.
> Remove the old RAM modules: Press the retaining clips to release them.
> Install the new RAM modules: Align the notches on the module with the notches on the slot and press down until the clips lock the RAM in place.
> Close the computer case, plug it in, and power it on.

c. Adding a Solid-State Drive (SSD) to Improve Laptop Performance:

Why: Replacing a traditional hard drive with an SSD can dramatically improve a laptop's speed and responsiveness.

Steps:

- ➢ Check compatibility: Ensure the laptop supports the SSD form factor (e.g., 2.5-inch or M.2) and interface (SATA or NVMe).
- ➢ Back up data: Create a complete backup of your laptop's data to avoid data loss during the upgrade.
- ➢ Turn off the laptop and unplug it.
- ➢ Open the laptop: Depending on the laptop model, this may involve removing the back panel or keyboard.
- ➢ Remove the old hard drive: Disconnect the data and power cables and unscrew it from its mounting.
- ➢ Install the SSD: Connect it using the same cables and screws, ensuring a secure fit.

- ➢ Reassemble the laptop and power it on.
- ➢ Install the operating system: You may need to install the operating system on the new SSD or clone the old drive to the new one.

d. Replacing a Smartphone Battery:

Why: Over time, smartphone batteries degrade, leading to shorter battery life. Replacing the battery can restore your phone's battery performance.

Steps:

- ➢ Purchase a compatible battery: Ensure the replacement battery is compatible with your smartphone model.
- ➢ Power off the smartphone and remove any protective case.
- ➢ Open the smartphone: Refer to the user manual or online tutorials for

instructions on accessing the battery.

➢ Disconnect the old battery: Gently detach the battery connector.

➢ Install the new battery: Connect it in place of the old one.

➢ Reassemble the smartphone and power it on.

These practical examples demonstrate how maintenance and upgrades can be performed on specific devices, enhancing their performance and extending their lifespan. However, always consult user manuals, manufacturer guidelines, and safety precautions when conducting these procedures, and consider seeking professional assistance for complex tasks if you're unsure.

Chapter 9: Safety and Best Practices

Chapter 9 is dedicated to safety considerations and best practices when working with electronics, ensuring that you are well-prepared to engage in DIY electronics repair responsibly and securely.

1. Advanced Safety Measures for High-Voltage Electronics

Working with high-voltage electronics presents significant risks that require advanced safety measures to protect yourself and your equipment. Here are key safety precautions for high-voltage electronics:

a. Isolation Transformers:

> ***Purpose***: Isolation transformers provide electrical isolation between the power source and the device you're working on. They protect you from electrical shock by breaking the direct electrical connection.

> ***Usage***: Connect the device to the output side of the isolation transformer, ensuring it's properly grounded. When using isolation transformers, even if a fault occurs, you're less likely to be exposed to the full line voltage.

b. Personal Protective Equipment (PPE):

➤ *Insulated Gloves*: Use high-voltage insulated gloves rated for the voltage you're working with. Make sure they're in good condition, free from cuts or punctures.

➤ *Safety Glasses*: Wear safety glasses with side shields to protect your eyes from potential electrical arcs and sparks.

➤ *Protective Clothing*: Use appropriate protective clothing, including flame-resistant, non-conductive, and insulating materials.

c. Controlled Environment:

> *Dedicated Workspace*: Set up a dedicated work area for high-voltage electronics. This space should be well-ventilated and equipped with safety equipment.

> *Safety Interlocks*: Install safety interlocks and emergency shut-off systems to disconnect power quickly in case of an emergency.

d. Safety Procedures:

> *Lockout-Tagout*: Implement a lockout-tagout (LOTO) system to de-energize and secure high-voltage equipment during maintenance or repair.

➤ **Risk Assessment**: Conduct a thorough risk assessment before working with high-voltage electronics. Identify potential hazards and plan for mitigating measures.

➤ **Buddy System**: When working with high-voltage systems, it's advisable to have a colleague or "safety buddy" nearby who can provide assistance and call for help in case of an emergency.

e. Emergency Response Plan:

➤ **Emergency Procedures**: Establish and document clear emergency procedures. This includes actions to take in case of electrical shock, fire, or other accidents.

➢ **First Aid Kit**: Ensure your workspace is equipped with a well-stocked first aid kit, and know how to use it.

f. Electrical Safety Training:

➢ **Education**: Ensure that anyone working with high-voltage electronics is properly trained in electrical safety procedures, understands potential hazards, and knows how to use safety equipment.

g. Grounding and Bonding:

➢ **Grounding**: Grounding of equipment is essential to protect against electrical faults. Ground all conductive materials and ensure a low-resistance path to ground.

8. Testing and Verification:

➢ *Voltage Measurement*: Always confirm that equipment is de-energized and safe to work on using appropriate voltage measurement devices. Use a voltage tester or multimeter to verify that no voltage is present.

9. Risk Assessment and Hazard Identification:

➢ *Hazard Analysis*: Conduct a thorough hazard analysis to identify potential electrical, mechanical, and environmental risks associated with high-voltage work.

10. Ongoing Monitoring:

➢ **Regular Inspections**: Periodically inspect safety equipment, personal protective gear, and workspaces to ensure they are in good condition and comply with safety standards.

These advanced safety measures are essential for working with high-voltage electronics. Remember that any work involving high-voltage systems should be conducted by trained and qualified personnel who are familiar with the specific equipment and safety protocols. Safety should always be the top priority when dealing with high-voltage electronics to prevent accidents, injuries, and damage to equipment.

2. Handling Electrostatic Discharge (ESD) in Electronics

Electrostatic discharge (ESD) poses significant risks to sensitive electronic components. It occurs when two objects with different electrical charges come into contact, leading to a sudden flow of electricity. ESD can cause permanent damage to microchips, transistors, and other electronic parts. Here's how to protect sensitive electronic components from ESD:

a. Understanding ESD Risks:

➢ **Component Damage**: ESD can cause microscopic damage to semiconductor devices, leading to latent defects that may result in early or intermittent failures.

➤ **Data Loss**: ESD can corrupt data stored in electronic devices, causing data loss.

b.ESD-Safe Workstations:

➤ **ESD-Protected Area**: Establish a designated ESD-protected workspace. This area should be equipped with ESD-safe materials, including ESD mats and wrist straps.

➤ **Workstation Grounding**: Connect ESD mats to a proper grounding point. This ensures that any static charge on your body is dissipated safely to the ground.

c. ESD-Safe Wrist Straps:

➢ **Wrist Straps**: Wear ESD wrist straps connected to an ESD-safe mat or grounding point on your workstation. These straps keep you at the same electrical potential as the electronic components you're handling.

➢ **Proper Connection**: Ensure the wrist strap makes good contact with your skin. The strap should fit snugly, with its metal plate in direct contact with your skin.

d. ESD-Safe Clothing:

➢ **Anti-Static Garments**: In environments with high ESD risk, consider wearing anti-static clothing. These garments are designed to minimize static buildup.

> **Avoid Synthetic Fabrics**: Avoid wearing clothing made of synthetic materials like polyester, which can generate static electricity. Cotton or other natural fibers are better choices.

e. ESD-Safe Tools and Accessories:

> **ESD-Safe Screwdrivers and Tweezers**: Use tools with ESD-safe handles to minimize the risk of static discharge through your tools.

> **Static-Dissipative Containers**: Store sensitive components and devices in static-dissipative

containers when they are not in use. These containers prevent the buildup of static charges.

f. ESD Awareness:

> *Training*: Ensure that personnel who handle sensitive electronic components receive training on ESD risks and safety measures.

> *ESD Signs*: Display ESD warning signs in areas where ESD-sensitive components are handled.

g. Proper Handling:

> *Minimize Movement*: Avoid unnecessary movement when working with sensitive components. Rapid or excessive movement can generate static charges.

➢ *Ground Yourself*: Before handling sensitive components, touch a grounded metal surface to discharge any static charge from your body.

h. ESD Control Materials:

➢ *Anti-Static Bags*: Store sensitive components in anti-static bags, which are specially designed to prevent static buildup.

i. Regular Inspections:

➢ **Equipment Checks**: Regularly inspect ESD equipment like wrist straps, mats, and grounding connections to ensure they are in good condition and functioning correctly.

Adhering to ESD safety protocols is essential to protect sensitive electronic components from damage. Implementing these precautions, whether in a manufacturing environment or during DIY electronic repairs, can prevent costly failures and data loss while preserving the integrity of electronic devices.

3. Emergency Response Procedures for Accidents in Electronics and Chemical Exposure

Accidents can happen when working with electronics or chemicals. It's crucial to be prepared and know how to respond in case of an emergency. Here are the general emergency response procedures, contacts, and resources:

A. In Case of Electric Shock:

➢ Ensure Safety: Ensure that the area is safe. If the victim is still in contact with the electrical source, do not touch them or the source until it is safe to do so.

➢ Call for Help: Dial emergency services (911 in the United States) to request immediate medical assistance.

➢ Power Off: If it's safe to do so, turn off the power source or unplug it. Use a non-conductive object, like a wooden stick, to move the victim away from the source if necessary.

➢ CPR and First Aid: If the victim is unresponsive or not breathing, initiate CPR if you are trained. If

not, perform basic first aid, such as checking for breathing and providing CPR if needed.
➤ Wait for Professionals: It's essential to wait for medical professionals to arrive for proper treatment and evaluation.

B. In Case of Chemical Exposure:

➤ Safety First: Ensure your safety. If you've been exposed to a chemical, move to a well-ventilated area and remove contaminated clothing.
➤ Call for Help: Dial emergency services to request immediate medical assistance or the Poison Control Center (in the United States, call 1-800-222-1222).
➤ First Aid: Follow any first-aid instructions provided for chemical

exposure on the product's label or Material Safety Data Sheet (MSDS).

➤ Decontamination: If instructed, rinse the affected area with copious amounts of water. Remove contaminated clothing and continue rinsing the skin or eyes for at least 15 minutes.

➤ Seek Medical Attention: Even if initial symptoms are mild, seek medical attention as some chemical exposures may have delayed or severe effects.

C. Contacts and Resources:

➤ Emergency Services: Know the emergency services number in your region (e.g., 911 in the United States) for immediate medical assistance.

- ➤ Poison Control Center: In the United States, you can contact the Poison Control Center at 1-800-222-1222 for guidance on chemical exposures.
- ➤ Safety Data Sheets (SDS/MSDS): These documents provide critical safety information about chemicals. Always consult the SDS/MSDS for any chemical you're working with.
- ➤ Local Hospital: Know the location and contact information for the nearest hospital or medical facility where you can seek emergency medical treatment.
- ➤ Safety Training: Consider taking courses in basic first aid and CPR, as these skills are valuable in emergency situations.

Remember, the best way to deal with accidents is to prevent them through safe work practices and the use of personal protective equipment. Regular training in safety procedures and maintaining a well-equipped first-aid kit can also be life-saving in emergency situations. If in doubt, always seek professional medical assistance.

4. Ethical Considerations in DIY Electronics Repair

DIY electronics repair offers numerous benefits, but it comes with ethical responsibilities that individuals should be mindful of. Here are some key ethical considerations:

A. Respecting Warranties:

Responsibility: When repairing a device, respect the manufacturer's warranty terms. Opening a device or making unauthorized repairs may void the warranty.

Transparency: If the device is under warranty, it's ethical to consider contacting the manufacturer or an authorized service center for repairs before attempting DIY fixes.

B. Avoiding Unauthorized Modifications:

Device Functionality: Unauthorized modifications can impact the device's functionality, safety, or warranty status. It's essential to refrain from making unapproved changes that may compromise these aspects.

Safety: Modifying a device without proper knowledge or authorization can lead to unsafe conditions, accidents, or legal issues. Ethical DIY repair should prioritize safety.

C. Adhering to Legal and Ethical Practices:

Copyright and Licensing: Respect intellectual property rights, copyrights, and software licenses when dealing with electronics. Avoid using unauthorized or pirated software in the repair process.

Data Privacy: When handling electronic devices, ensure that any data present on the device is treated with confidentiality and privacy. Ethical repair involves safeguarding personal and sensitive information.

Environmental Responsibility: Ethical DIY electronics repair should include proper disposal of electronic waste, recycling when possible, and adherence to environmental regulations.

D. Honesty and Transparency:

Documentation: Keep clear records of your repair process, including any changes made to the device. Honesty and transparency in documenting the repair can be crucial, especially if the device is later serviced by professionals.

Disclosing Previous Repairs: If you intend to sell a repaired device, be transparent about its history, including any prior DIY repairs. This maintains ethical and fair consumer practices.

E. Informed Consent:

Client's Involvement: If you're repairing a device for someone else, ensure they are fully informed about the repair process, potential risks, and any implications for warranties. Their informed consent is essential.

Quality Assurance: Guarantee that your repair work meets industry standards and safety practices to ensure the repaired device functions safely and effectively.

F. Skills and Knowledge:

Continuous Learning: Strive to enhance your repair skills and knowledge, and be aware of your limitations. Avoid taking on projects that exceed your expertise, as this could lead to poor repair work or damage to devices.

By adhering to these ethical considerations in DIY electronics repair, you contribute to a culture of responsibility, transparency, and safety. Furthermore, ethical DIY repair practices help maintain the trust and integrity of the repair community, manufacturers, and consumers alike.

5. Guidance for Working on Specific Types of Electronic Devices:

Each type of electronic device presents unique risks and challenges when it comes to repair. Here's tailored safety guidance for common categories of electronic devices:

A. Computers and Laptops:

> ➤ Electrical Hazards: Unplug the computer or laptop and ensure it's powered off before opening the case. Be cautious around power supplies and capacitors.
> ➤ Static Discharge: Use anti-static precautions like wrist straps and mats to prevent damage to sensitive components.
> ➤ Heat Management: Be aware of hot components like processors and power supplies. Wait for them to cool down if necessary.

B. Smartphones and Tablets:

> ➤ Glass Safety: When replacing screens, be extremely cautious

with broken glass. Wear protective gloves and eye protection.

> Battery Hazards: Handle lithium-ion batteries with care. Avoid puncturing or short-circuiting them, as they can catch fire or explode.

> Small Components: Small screws, connectors, and components can be easily lost. Use an organized workspace and small containers to store parts.

C. Home Appliances:

> Electrical Hazards: Unplug the appliance from the power source before beginning any repair. Ensure that it's properly grounded.

> Gas Appliances: For gas appliances, shut off the gas supply and ensure proper ventilation to prevent gas leaks.

➤ Water Damage: For appliances involving water (e.g., washing machines), be cautious of electrical connections near water sources.

D. Audio Equipment:

➤ Amplifiers: Large amplifiers can store dangerous electrical charges, even when unplugged. Discharge capacitors safely.
➤ Loudspeakers: Avoid exposure to high sound levels, especially when testing speakers.
➤ Cables: Inspect and replace frayed or damaged cables, as they can pose electrical and fire hazards.

E. Gaming Consoles:

➤ Ventilation: Ensure gaming consoles have proper ventilation to prevent overheating and component damage.

➤ Optical Drives: Be cautious when repairing optical drives, as laser diodes can damage eyesight.

➤ Power Supplies: When repairing power supplies, watch out for capacitors that may retain a charge even when unplugged.

F. Televisions and Monitors:

➤ CRT Monitors: Cathode ray tube (CRT) monitors can store high voltages even when turned off. Use extreme caution.

➤ LCD and LED Displays: Be cautious when handling screens, as they are fragile and can be easily damaged.

➤ High Voltage: Some TVs and monitors have high-voltage

components. Exercise caution when working on these parts.

Remember, the key to safe repairs is understanding the unique risks associated with each type of electronic device and taking appropriate precautions. Always consult the device's manual and follow manufacturer-specific safety guidelines. If you're unsure about a specific repair, consider seeking professional assistance.

6. Responsible Disposal of Electronic Waste:

Responsible disposal of electronic waste (e-waste) is crucial for both environmental preservation and public health. Here are some options for recycling or disposing of old and non-repairable electronic devices:

A. E-Waste Recycling Centers:

- ➢ Drop-Off Locations: Many cities and regions have dedicated e-waste recycling centers. Check your local listings for the nearest drop-off location.
- ➢ Accepted Items: E-waste recycling centers typically accept a wide range of electronics, including old computers, TVs, smartphones, and more.
- ➢ Data Security: Before recycling, ensure you've wiped all personal data from devices or removed the storage components to protect your privacy.

B. Manufacturer Take-Back Programs:

> ➤ Manufacturer Initiatives: Several electronics manufacturers have take-back programs, where they accept their old products for recycling. Check with the manufacturer of your device to see if they offer this service.

C. Retailer Recycling Programs:

> ➤ Retailer Initiatives: Some electronics retailers provide e-waste recycling services. They may take back old devices, especially when you're purchasing a new one.
> ➤ Trade-In Programs: Some retailers also offer trade-in programs, allowing you to exchange your old device for credit towards a new purchase.

D. Local E-Waste Collection Events:

➤ Community Events: Many communities host periodic e-waste collection events where you can drop off your old electronics for recycling.

E. Certified E-Waste Recyclers:

➤ Certifications: Look for certified e-waste recycling facilities. Organizations like R2 (Responsible Recycling) and e-Stewards ensure that electronics are recycled responsibly.

F. Mail-In Recycling Programs:

➤ Mail-In Services: Some companies offer mail-in recycling services. They provide you with packaging to send your old electronics for recycling.

G. Donate or Repurpose:

- ➢ Functional Devices: If your device is still functional, consider donating it to a charitable organization, school, or someone in need.
- ➢ Creative Reuse: Explore creative ways to repurpose old devices for projects or as dedicated home automation controllers.

H. Trade-In or Resell:

- ➢ Marketplace Platforms: You can also sell or trade in your old devices on platforms like eBay or through smartphone trade-in programs.

I. Hazardous Waste Collection Programs:

➤ Special Consideration: For items with hazardous components like batteries or certain chemicals, check with your local hazardous waste collection programs for safe disposal options.

J. Proper Battery Disposal:

➤ Battery Recycling: Batteries from electronic devices can be recycled separately. Many retailers and recycling centers accept batteries.

K. Legislative Compliance:

➤ Laws and Regulations: Be aware of local and national e-waste disposal regulations. Many countries and regions have laws governing the responsible disposal of electronic waste.

By following these responsible e-waste disposal practices, you contribute to environmental sustainability and help prevent hazardous materials from entering landfills. Proper disposal is not only ethical but essential for the preservation of our planet.

Overall, "DIY Electronics Repair at Home" is a comprehensive, well-structured guide that empowers you with the knowledge and practical skills to diagnose, repair, and maintain electronic devices in their homes. It covers a wide range of topics, ensuring you have the tools and understanding needed to tackle a variety of common issues in electronics, all while prioritizing safety and responsible practices.

THANK YOU